T0357902

THERMODYNAMICS
AND
FLUID MECHANICS

AN INTRODUCTION

THERMODYNAMICS AND FLUID MECHANICS

AN INTRODUCTION

ROGER KINSKY

The McGraw-Hill Companies, Inc.

Beijing Bogotà Boston Burr Ridge IL Caracas
Dubuque IA Lisbon London Madison WI
Madrid Mexico City Milan Montreal New Delhi
New York San Francisco Santiago Seoul
Singapore St Louis Sydney Taipei Toronto

Reprinted 1996, 1997, 2000, 2001, 2004 (twice), 2006 (twice), 2007, 2008 , 2010 , 2024

National Library of Australia Cataloguing-in-Publication data:

    Kinsky, Roger.
        Introductory thermodynamics and fluid mechanics.

        Includes index.
        ISBN 10: 0 07 470238 6
        ISBN 13: 978 0 07 470238 3

        1. Thermodynamics 2. Fluid mechanics. I. Title.
536.7

Published in Australia by
McGraw-Hill Australia Pty Limited
Level 33 World Square, 680 George Street Sydney, NSW 2000, Australia

Acquisitions Editor: Nicola Cowdroy
Production Editors: Caroline Hunter and Marjorie Pressley
Designed and Illustrated by :   George Sirett

Typeset in 10/11.5 Times
in Australia by Monoset Typesetters, Brisbane
Printed in Australia by Pegasus

Contents

≡

List of Tables

Preface

This book was written in response to the changes that have occurred as a result of governmental action in Australia to standardise the accreditation, content and assessment of engineering courses at a trade and post-trade level. The main thrust of this was the introduction of a number of standard modules of duration approximately 40 hours each.

This book covers two such modules, namely: Thermodynamics 1 and Fluid Mechanics 1. Thermodynamics 1 is dealt with in Part 1 (Chapters 1–7) and Fluid Mechanics 1 is dealt with in Part 2 (Chapters 8–13).

The decision to incorporate two modules into one book was made on essentially economic grounds and provides best value for students who now need to purchase only one book rather than two in order to study both modules. A spin-off benefit is that there is material common to both modules, which did not need to be duplicated. For example, basic concepts such as mass, volume, pressure and such like are common to both thermodynamics and fluid mechanics as well as basic laws and principles such as the conservation of mass law (leading to the continuity equation), the conservation of energy law, gas laws and others.

Each part of the book is essentially 'stand alone'; however, in order to avoid duplication of the same material in both parts, some cross-referencing is necessary from Part 2 to Part 1. For example, gas laws and processes are dealt with in Part 1 and are not repeated in Part 2.

The initial intention was essentially to 'cut and paste' from my existing books *Heat Engineering* and *Applied Fluid Mechanics*. However, as work proceeded I found it necessary to re-write much of the material, so although there are many similarities, there are also many differences. I made every effort to ensure that the differences were an improvement on the original and to make the book as 'user friendly' as possible. The main features of this book are as follows:

- The only pre-requisite knowledge and skills required involve basic mathematics, language, and engineering science.
- Each chapter covers a discrete section in the relevant module, with essentially the same learning outcomes and objectives. Also the sequencing of the chapters and the material presented closely follow the relevant module.
- Short sentences and paragraphs and point-by-point format are used throughout. Meaning is conveyed using the simplest possible language and with the avoidance of unfamiliar terminology or jargon wherever this could be done without sacrificing precision of expression.
- Symbols have been simplified and standardised. The meaning of each symbol and the correct units for it are stated whenever a new symbol is introduced. In addition, a list of principal symbols with quantity and units is given in Appendix 1.
- The mathematics has been simplified and equation derivations are omitted in the text. However, some principal equation derivations are given in Appendixes 5–13. In the

text, each important equation is boxed, numbered and given a descriptive title. Also, the principal equations are listed in Appendix 2. To simplify worked examples, the symbol $A(d)$ is used rather than the more complex $\pi d^2/4$ for the area of a circle of diameter d.

- There is a list of objectives at the beginning of each chapter which closely match those given in the relevant module. There is also a summary of the important points at the conclusion of each chapter.
- Numerous worked examples and self-test problems are given throughout. The solution to each self-test problem is given in a section at the end of the book. These self-test problems were incorporated to assist in making the book 'user friendly' for both students and teachers. Students will obtain maximum benefit from these self-test problems if they do not refer to the solution given until completion of their own workings.
- Each chapter has an appropriate number of solved problems graded in approximate order of difficulty.
- A Teachers Manual containing solutions to the problems is available free to adopters of the text.

While every attempt has been made to produce a book free of errors, the practicality is that some errors will no doubt appear. I would be most grateful to have these brought to my attention. I would also be pleased to hear from any teachers or students who have any comments or suggestions for improvement.

ROGER KINSKY

Sydney, December 1994

PART 1

Thermodynamics

Energy and humanity

Introduction

Thermodynamics is the study of heat energy or thermal energy and the processes by which this energy may be converted into other forms of energy. Engineers need a knowledge of thermodynamics because almost all energy used by humans derives directly or indirectly from thermal energy.

This chapter treats energy and energy-conversion processes in a primarily descriptive way. Quantitative methods are given in greater detail in later chapters.

1.1 THE NEED FOR ENERGY

Energy is needed to sustain and enrich human life. Typical energy uses are listed in Table 1.1.

Table 1.1 *Typical energy uses*

Type of energy	Uses
heating	• human comfort • cooking and food preservation • Industrial processes requiring heat such as metal and oil refining, glassmaking etc.
cooling	• human comfort • food storage and preservation • industrial processes requiring cooling such as cryogenics, environment control in computer rooms etc.
mechanical	• human transportation • materials movement • raw materials extraction • mixing and stirring • mechanical processes such as metal forming and cutting • pumping and compressing fluids
light and sound	• human comfort and communication
electricity	• industrial processes such as welding and plating • electronic/electrical devices such as computers • conversion to other energy forms, for example with an electric motor

Notes

1   Energy is measured in joules (J). Often the rate of transfer (or conversion) of energy is important. This is measured in joules per second (J/s) or watts (W).

2   Mechanical energy includes potential energy, kinetic energy and work. Work transfer rate is known as **power**. The power of a rotating shaft is also known as **shaft power**.

3   The mechanical energy of a fluid is also known as **fluid power**.

The productivity of a nation is dependent on the output of goods and services by each inhabitant. This reflects the standard of living (by modern standards) and may be measured by the gross national product (GNP) per capita or the gross domestic product (GDP) per capita. A graph of energy consumption per capita against productivity is given in Figure 1.1 and this shows clearly that high productivity (and standard of living) requires a high usage of energy.

1.2   *ENERGY CONVERSION*

Energy exists in many forms and often only one form is suitable for a particular purpose. For example, human muscle power is derived from chemical energy in food that is eaten, and warmth may be obtained from solar heat. However, the human body cannot directly convert solar energy into muscle power. A conversion process is necessary—

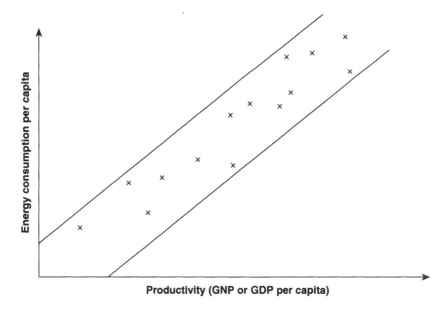

Fig. 1.1   *Energy consumption per capita correlated with productivity (GNP or GDP per capita)*

photosynthesis. *This process utilises solar energy to convert simple substances into more complex compounds, hence storing chemical energy.* The chemical energy is subsequently released after digestion of the food. Electrical or electronic equipment is powered by electricity. Electrical energy occurs in nature as lightning, but in this form it cannot easily be tapped into or stored, so the electrical energy needed by humans must be obtained by conversion from some other energy source.

Some of the most significant energy-conversion processes are listed in Table 1.2.

Table 1.2   *Significant energy-conversion processes*

Conversion process	Conversion mechanism
mechanical → heat	friction (mechanical or fluid)
heat → mechanical	heat engine
heat → electrical	MHD (magnetohydrodynamics)
mechanical → electrical	electric generator
mechanical → fluid power	pump, compressor, fan
fluid power → mechanical	turbine, fluid motor, wind vane, wave vane
electrical → heat	induction or resistance heater
electrical → mechanical	electric motor (rotary or linear)
electrical → chemical	electrolysis, or charging storage battery
solar → heat	solar collector or absorber
solar → electrical	solar cell (photo-voltaic cell)
solar → chemical	photosynthesis or photolysis
chemical → heat	combustion and food digestion
chemical → electrical	fuel cell, or discharging storage battery
nuclear → heat	nuclear reactor

In many cases, several conversion steps may be necessary. For example, in nature, solar energy is converted to mechanical energy in wind, waves and rain. Each of these may be converted into electrical energy, for example with a wind generator or a water turbine. A more complex series of conversion steps utilising electrical energy is shown in Figure 1.2.

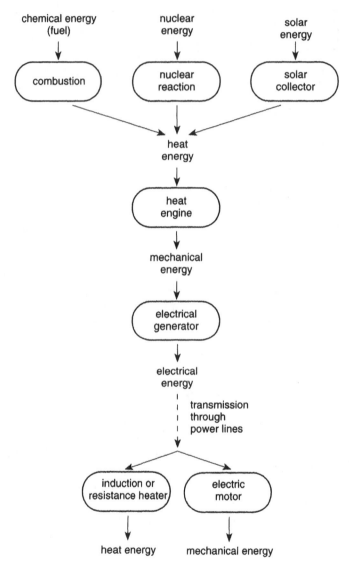

Fig. 1.2   *Typical energy-conversion processes*

Efficiency of conversion

Most conversion steps involve a loss of usable energy. The efficiency of conversion is most important and is given by the following equation:

$$\eta = \frac{\text{usable energy output}}{\text{energy input}}$$

Typical energy-conversion efficiencies are given in Table 1.3.

Table 1.3 *Typical energy-conversion efficiencies*

Means of conversion	Efficiency (%)
electrical-resistance heater	100
combustion heater	100 (non-vented flue)
	80 (vented flue)
electric generator (500 MW)	98
hydraulic turbine (50 MW)	94
electric motor (250 kW)	94
(1 kW)	78
storage battery	74
fuel cell	60
diesel engine (50 MW)	40–45
(50 kW)	35–40
petrol engine (50 kW)	20–35
steam power station	35–40
solar cell	10–20
steam engine (locomotive)	5–10
photosynthesis	0.4–3

Direct conversion

With consecutive conversion steps, *the total efficiency is the product of the individual efficiencies*. For example, with four steps, each with an efficiency of 90%, the overall efficiency is $0.9 \times 0.9 \times 0.9 \times 0.9 = 0.6561 \doteqdot 65\%$. For this reason it is desirable to minimise the number of conversion steps. In direct conversion, energy is converted directly from source form to consumed form without any intermediate steps. Although in principle this makes for maximum efficiency, in practice the feasibility of direct conversion depends on both the cost and the efficiency of the device. For example, the use of solar cells for generation of electricity is limited because of their high cost and low efficiency. Similarly, it is possible to directly convert solar energy into mechanical energy, but the efficiency is so low that such devices are useful only as demonstration models.*

1.3 *HEAT ENGINES*

Solar energy, combustion of a fuel, and nuclear reactions all produce heat. Apart from heat itself, the most needed form of energy for humans is mechanical energy. The device that converts heat energy to mechanical energy continuously is the **heat engine**, hence the heat engine is of special significance. Indeed the development of the heat engine brought about the Industrial Revolution and the emergence of the modern way of life.

---

* In one such device, a paddle with different metallic surfaces on each side is suspended in an evacuated glass vessel. The paddle rotates when the device is exposed to radiant energy because of the different absorption and emission rates of the two surfaces.

The theory of heat engines is discussed in Chapter 6, where it is shown that there are special limitations to the maximum efficiency attainable by a heat engine. Early heat engines were very inefficient but, nowadays, engines such as large diesel engines used in the propulsion of ships can have efficiencies of 40–45% and steam power stations have efficiencies of 35–40%.

1.4   AVAILABILITY OF HEAT ENERGY

Energy cannot be created (or destroyed) but can be converted from one form to another or from one level of availability to another. For example, when a fuel is burnt, x J of chemical energy is converted into x J of high-temperature heat energy. When this heat is used to warm a room or a person, x J of heat at a high temperature becomes x J of heat at a low temperature. The quantity of heat, x J, at a low temperature is no longer useful. It eventually warms the atmosphere and is radiated into space.

The **availability** (or usefulness) of heat energy (known also as the **grade** of the heat energy) is measured by its **entropy**; the lower the entropy, the more useful the energy (i.e. the higher the grade). Entropy is a difficult concept to understand, mainly because the human body cannot sense it. Basically, *entropy is the quantity of heat divided by absolute temperature*; hence the higher the temperature at which heat occurs, the lower the entropy. What really occurs when heat energy is 'used' is that the entropy increases and the heat energy becomes less useful.

The availability or usefulness of heat energy (the entropy) is also directly related to the efficiency with which it can be converted to mechanical energy. For example, if a source of energy provides heat at 2000°C, the maximum theoretical conversion efficiency is 87%. However, if the same quantity of heat is at source temperature 40°C, the efficiency is only 6.4%. For a given power output, the amount of energy needed, and the capital and running costs involved in energy conversion, increase as the efficiency falls. At an efficiency of 6.4%, the energy input and the size of plant need to be fourteen times more than if the efficiency is 87%.

Hence, as heat temperature falls and entropy rises, the cost of energy conversion increases to the point where the conversion may no longer be viable.

Self-test problem 1.1

(a) What is meant by the *efficiency of energy conversion*?

(b) State *two* reasons why it is important to obtain as high a conversion efficiency as possible.

(c) What is meant by *direct conversion of energy*, and why may it be advantageous?

(d) If technology could continually make improvements, would it be possible to eventually obtain 100% efficiency with every energy-conversion device? If not, for which device or devices would this not be possible?

(e) Three energy-conversion devices each has an efficiency of 80%. What is the overall efficiency of energy conversion if these devices are connected:
   (i) in parallel?
   (ii) in series?

1.5   SOURCES OF ENERGY

Because energy cannot be created, all energy is derived from some source. The available sources of energy are shown in Figure 1.3.

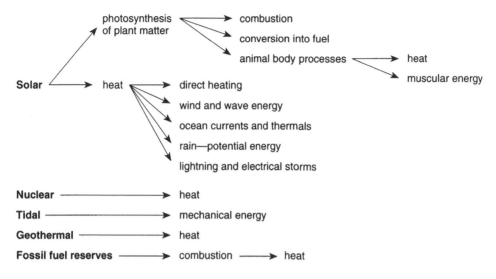

Fig. 1.3 *Sources of energy*

Notes

1 All the energy available to humans has a common source, namely nuclear energy. Solar energy is a result of nuclear reactions in the sun, and tidal energy is the result of past nuclear reactions when the earth was formed. Combustible fuels derive energy from stored solar energy and thus from nuclear energy.

2 Energy that is available continuously is known as renewable energy and consists essentially of solar energy, tidal energy and geothermal energy. Energy that derives from a source that cannot be re-created is known as non-renewable (or stored energy) and consists essentially of nuclear-fission energy and stored fuel reserves.

1.6 *SOLAR ENERGY—PHOTOSYNTHESIS*

Photosynthesis is the process by which plants convert solar energy into chemical energy. The plants use solar energy to drive the chemical reactions that produce cell material of complex chemical structure (known also as biomass) from basic elements such as carbon, oxygen, hydrogen and nitrogen. These elements are obtained from the earth, the atmosphere and water. A reverse process occurs during combustion of biomass, which converts chemical energy into heat energy. Many animals are able to digest plant material and convert chemical energy into muscular energy (mechanical energy) and into heat energy (in the case of humans and other warm-blooded animals) by a process that is essentially one of slow internal combustion.

Burning biomass material may present some difficulties because of moisture content and production of pollutants such as smoke, but these difficulties may be overcome by appropriate furnace design and pollution control equipment. Woodchips, bagasse (sugar-cane stalks after crushing), rice husks, wheat husks and similar organic materials can be burnt successfully and provide useful amounts of energy. An alternative is to convert biomass into a liquid or gaseous fuel. For example, alcohol, which is a clean-burning fuel suitable for use in motor vehicles, can be obtained from sugar cane or sugar beet by a fermentation and distillation process. Biomass may also be converted by bacteria into a clean burning gas (mostly methane). Waste biomass in garbage dumps, sewage plants, animal feed lots, dairies and so on can be used for this purpose.

The conversion of solar energy into chemical energy by photosynthesis may be of future value in Australia because we have sufficient land, with appropriate climate, to produce enough fuel by this method to have a significant impact on our energy requirements.

1.7   *SOLAR RADIATION*

The radiant energy received by the earth outside the atmosphere is about 1.4 kW/m². Due to the effect of the earth's atmosphere, much of the radiation is filtered, primarily by the ozone and water vapour, so that the amount of radiation received even on a prefectly clear day is only about 850 W/m² on the earth's surface. In addition, clouds and dust in the atmosphere can obscure the radiation, reducing it to a fraction of this value on overcast days.

The radiation received also depends on latitude. In northern latitudes (where much of the world's energy consumption takes place), the energy received per square metre on the surface of the earth is much less than that received in tropical regions, or in the centre of Australia. The energy received also varies throughout the year with the relative position of the sun in the sky overhead (thus accounting for the seasons).

A typical *mean* value during daylight hours in Australia is about 400 W/m². This is a relatively low energy density, and in order to obtain reasonable amounts of energy, large receiving areas are required.

Solar Collectors

Two types of solar collectors are available, namely **flat-plate collectors** and **focusing collectors**. A typical flat-plate collector used for water heating is illustrated in Figure 1.4.

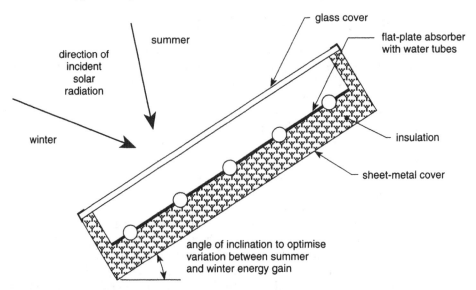

Fig. 1.4   *A typical flat-plate solar collector*

If higher temperatures are required than those obtainable with a flat-plate collector, a focusing collector may be used. This collector is illustrated in Figure 1.5. The high temperatures obtained with such a collector make solar heat more feasible as an energy source for a heat engine or refrigeration plant. However, focusing collectors require a complex and expensive tracking system, which severely limits their widespread application.

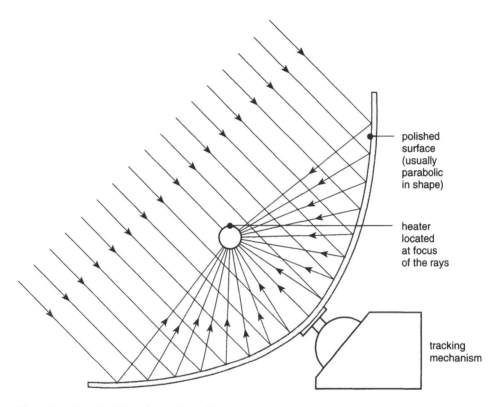

Fig. 1.5   *A typical focusing solar collector*

Solar heating or cooling

At the present time, the main use of solar energy for heating is in hot-water heaters. Other possibilities are space heating or cooling applications for home or industrial premises, but the problem here is that most heat is available in summer when it is usually not required, in fact *cooling* is desirable. Part of the solution can be incorporated in the architectural design of the building to take advantage of the sun in winter and screen the sun in summer. This may be augmented by the use of solar collectors, for example, to circulate warm water through the building in winter. In summer, the water may be cooled at night and circulated during the day.

Other feasible alternatives are to use solar-energy-assisted heat pumps for heating applications, or solar-driven absorption refrigeration systems for cooling applications.

Solar cells

Solar cells (or photo-voltaic cells) convert radiant solar energy directly into electrical energy. The conversion efficiency depends on the type of cell (and the cost). Cheaper cells have efficiencies of only about 10% but more costly cells may have an efficiency of 20%. Efficiencies of up to 25% are achievable with special cells but these are so expensive that they are of little current commercial value. If a cell has an efficiency of 10%, then a cell area of 25 m² is needed to produce an average of 1 kW of electricity during daylight hours. This cell area is too large and the cells too costly to make this a feasible method for large-scale power production. However, solar cells are ideal for low-power applications such as lighting of marine pylons or auxiliary power on pleasure craft or on farms in remote locations. In applications such as these, storage batteries usually provide peak power demands and solar cells are used to recharge the batteries over longer periods of time.

Power tower

The power-tower method of converting solar to electrical energy involves a series of mirrors located around a central tower, which houses a power-generation plant of the conventional type (Fig. 1.6).

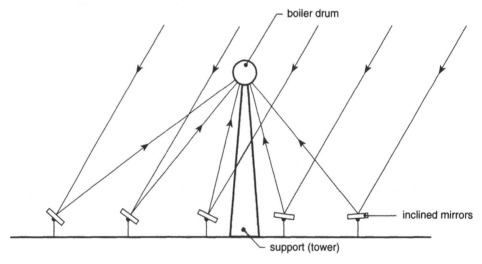

Fig. 1.6   *A power tower*

Each mirror must be inclined at a slightly different angle so that they all have a common focus on the boiler drum. Because of the motion of the sun in the sky, the angle needs to be constantly adjusted during the day; this requires sophisticated (and costly) equipment.

A conventional power-generating station usually has about a 1000 MW capacity. Hence if a power tower of the same capacity were located at a place on the earth's surface with average radiation 500 W/m², a collection area of 6×10^6 m² would be required, assuming an overall efficiency of 33%. This area is not impossibly large and would be provided by a circle of diameter 2.8 km. The main cost is in the mirrors themselves and in the equipment necessary to regulate their position.

This method is of proven feasibility and a plant large enough to supply the energy needs of a town has been constructed and is operating in the USA. For large-scale power production, the main difficulty (apart from cost) is that the best areas for receiving solar energy are in desert regions, in which the air is clear and little cloud occurs. The cost of *transmitting* the energy then becomes a major factor, and over long distances, it becomes uneconomic to use power-transmission lines due to the high capital cost and the energy loss.

Hydrogen from solar energy

The loss (and cost) associated with long electrical-transmission lines can be avoided if the solar-derived electrical energy is converted into chemical energy. One solution is to use electrolysis to convert water into hydrogen,* which can then be transferred through a pipeline as a gas, or be liquefied and transported by tanker. The cost of the associated plant and equipment is high, and the overall efficiency is low. However, the low efficiency may be partly offset in the reconversion process of hydrogen into mechanical energy because it is not essential that a heat engine be used (although this is possible). The alternative is to use a **fuel cell**, a device rather like a battery with an electrolyte such as potassium hydroxide. Hydrogen and oxygen enter through porous carbon tubes and combine to form water, and at the same time generate electricity. The electricity may be used to drive an electric motor.

By using a fuel cell, the efficiency of conversion of hydrogen into mechanical energy is about twice that of a heat engine.

The various energy-conversion processes are listed in Figure 1.7 (on page 14).

Ocean thermal-energy collector (OTEC)

The oceans of the world form large natural collection areas for solar energy. Since the solar energy is received at the surface, it does not convect through the water as would be the case with heating from below. Hence there is a temperature gradient in sea water. In tropical regions away from the influence of strong currents, the difference in temperature at the surface and at a depth of 100 m may be of the order of 15°C (with temperature differences up to 25°C possible at greater depths).

The idea of using this natural temperature gradient in order to operate a heat engine was conceived in 1929 by the Frenchman Georges Claude, who produced a 22 kW power plant operating off the coast of Cuba. With support from the French government, a 7.5 MW unit was subsequently built.

---

* Electrolysis is the process by which direct current is fed to two electrodes spaced at some distance apart in a water bath to which a conducting medium such as salt has been added. The current passing through the water from one electrode to the other causes the water to dissociate into hydrogen and oxygen.

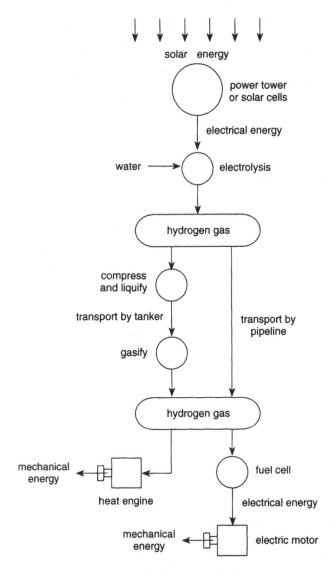

Fig. 1.7   *Energy-conversion processes from solar to mechanical energy using hydrogen*

These plants operated on an open cycle using sea water as the working fluid and proved the technical feasibility of this type of collector (the OTEC). However, further work was abandoned because the plants were not economically viable. Recent proposals are based on a closed cycle using propane or ammonia as the working substance, as illustrated in Figure 1.8. However, due to the comparatively small temperature difference between the hot and cold fluids, the ideal (Carnot) efficiency is low (<4%) and the actual efficiency is considerably lower.

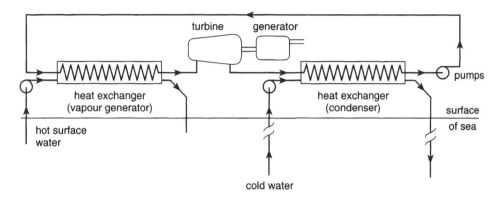

Fig. 1.8   *An ocean thermal-energy collector*

Solar ponds

Solar ponds are used in essentially the same way as the OTEC except that the temperature difference is produced in a salt-water pond of shallow depth and the hot water is at the bottom and not at the top. In the ocean or in a natural lake or pond, hot water is at the surface because hot water is less dense than cold water. At the surface, high heat dissipation to the atmosphere occurs, particularly at night or when a wind is blowing over the surface so that a high temperature gradient does not occur. The dissipative heat loss can be considerably reduced if the hot water is at the bottom rather than at the top because water above acts as a very effective insulator so that high temperature gradients are possible.

In order to prevent the hot water from rising to the top of the pond, the salinity of the pond is controlled during filling so that there is high-salinity (high-density) water at the top and low-salinity (low-density) water at the bottom. The pond is carefully filled from the top as the salinity is increased to ensure that no mixing occurs between the layers.

Like the OTEC, solar ponds have proven feasibility; for example, they are used in Israel for power production. However, despite the higher temperature difference, efficiency is still too low to make them economically viable except in special applications.

1.8   *WIND AND WAVE ENERGY*

Wind and wave energy are examples of indirect solar energy, that is, solar energy that has already been converted by nature to kinetic energy.

Windmills have been used for many centuries for grinding grain or pumping water. The power developed by a wind generator is proportional to the cube of the wind velocity, and at velocities over about 24 km/h, wind generators become economically viable. However, in most populated areas, the wind seldom reaches this speed and is spasmodic, so continuous power generation is not possible. One possibility is to site wind generators in locations where the wind blows with a high velocity for the majority of the year. The electrical energy could be used to manufacture hydrogen, which could then be transported to the usage point (see Fig. 1.7). The use of hydrogen is still in the experimental stage; however, wind farms and large wind generators have been constructed in many countries (including Australia) and are technically feasible but costly.

Another possibility for wind power is using sailing ships for transportation. However, the longer passage times required by sailing ships with large crews make them uneconomic compared with motor-driven vessels. As fuel costs inevitably rise, it is possible that power-assisted sailing ships with mechanically adjusted sails will become viable. Vessels of this type have been constructed and fuel savings of 10–15% have been reported.

Wave energy can also be converted to electrical energy by utilising the kinetic energy of waves to drive an electric generator. The amount of power generated depends on the size and speed of the waves, which vary with location and weather conditions. Studies suggest that useful amounts of power could be generated in coastal areas of Australia and in overseas locations such as Japan and the United Kingdom.

Many possible mechanical configurations have been investigated, such as a series of swivelling pontoons, which oscillate as the waves pass under them. This motion is then converted into rotary motion and drives an electric generator. Another method involves floating cylindrical buoys anchored to the sea floor so that their position remains fixed. Inverted floating cylinders are positioned around the fixed buoys and these rise and fall in response to wave motion. Using valves, air can be periodically inspirated, compressed and then delivered, the whole device acting, in effect, as a large air compressor. The compressed air can then be used to drive an air motor or turbine coupled to an electric generator.

There are some mechanical difficulties associated with these proposals, such as the destructive effect of storms, the corrosiveness of sea water and the accumulation of marine growths on surfaces. These problems are not insurmountable, but again economics is the chief factor in determining future viability.

1.9 HYDROELECTRIC POWER

Another application of the indirect use of solar energy is that of hydroelectric power. Solar heat evaporates water, which rises as vapour, condenses in clouds and eventually falls back to the earth as rain. If the rain water is collected in a dam at some high elevation, the potential energy of the water may then be used to drive a hydraulic turbine coupled to an electric generator. The method is of proven feasibility (turbine–generator efficiency is about 90%) and many hydroelectric power stations are in operation throughout the world. In Australia, the total full hydroelectric generating capacity is 6600 MW, which is about 20% of total requirement. However, there is insufficient water to enable continuous operation at full capacity; the contribution of hydroelectricity to total electricity generation averages less than 4%. Water turbines (unlike steam turbines) can be easily started up and shut down and hence hydroelectric power is more useful for providing peak power demand.

Another advantage of hydroelectric power is that of reverse operation. The turbines and generators may be constructed so they will operate in reverse and pump water back up into the dam. This enables the storage of energy from any variable source and the smoothing out of peaks in power demand.

1.10 NUCLEAR ENERGY

There are three types of nuclear reactors, namely **fission reactors** (slow), **breeder reactors** (fast) and **fusion reactors**.

Fission reactors

The simplest type of fission reactor is called a **burner reactor**, because the fuel is used up in a similar way to that in which a fossil fuel is burnt in a conventional power station. There is a wide variety of reactor designs and materials; a typical basic design is illustrated schematically in Figure 1.9.

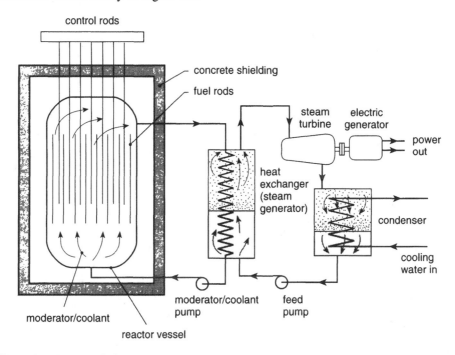

Fig. 1.9 *A typical fission-reactor power-generation plant*

The reactor coupled to a heat exchanger takes the place of the coal-fired steam generator, otherwise the downstream end is the same as in a coal-fired power station. The principal components of the reactor are as follows:

Fuel   Only three fissionable fuels are suitable, namely U^{235}, U^{233} (uranium) or Pu^{239} (plutonium). Currently, the large majority of nuclear reactors use U^{235} as the fuel. Since U^{235} exists in natural uranium which is (U^{238}) as only 0.71%, it is common to enrich the U^{238} by processing it first to increase the proportion of U^{235}. This is then known as *enriched* fuel, which is generally processed into rod, tube or plate form.

Control rods   These are used to control the multiplication factor of neutron production. If this is *greater* than 1, the reaction will get out of control, and if *less* than 1, the reaction will slow down and may stop. Cadmium and boron are often used because they have a great capacity for absorbing neutrons.

Moderator/coolant   The function of the moderator is to absorb some of the neutron energy in order to make the fissioning most effective. Water, graphite or deuterium oxide are most often used. The coolant has the function of absorbing the energy of the reaction. In many cases, the moderator also serves as the coolant; however, in some reactor designs, different moderating and coolant materials are used.

Shielding   Usually a thick concrete shield surrounds the reactor in order to absorb the radiation that accompanies the reaction.

The technology for producing energy with fission reactors is well proven and there are over 400 such reactors currently operating throughout the world. Costs compare favourably with coal-fired power stations and there is capacity to provide energy well into the future. No waste products are discharged to the atmosphere, and uranium mining is less hazardous than coal mining. Against these advantages, there is the problem of nuclear-waste disposal and the possibility of catastrophic mishap such as occurred at Chernobyl in the USSR in 1986.

In Australia, nuclear energy is not used for power production because of its vast coal reserves and the public opposition to nuclear energy.

Breeder reactors

Uranium occurs in nature as uranium oxide, U_3O_8. It is found throughout the earth's crust and, indeed, also in sea water in varying concentrations. Like gold or any other mineral, its extraction is only feasible if the reserves are rich enough to make extraction economic. Australia is particularly fortunate in having vast high-grade deposits of uranium.

However, since U^{235} is only 0.71% of natural uranium (U^{238}), rising world energy consumption would quickly deplete the economic uranium reserves. A far longer term solution is available by means of stepwise reactions. If U^{238} is made to capture a neutron, it becomes U^{239}; however, this is unstable and decays quickly to Np^{239} (neptunium), which is also unstable and decays to Pu^{239}, which is fissile and may be used as a reactor fuel.

Reactors that produce more fissile material than they consume are called **breeder reactors**. They do this by dispensing with the moderator so that they are also called 'fast reactors'. Instead of absorbing the excess neutrons produced by the fission of the U^{235}, these neutrons are used to inititate the conversion of U^{238} to Pu^{239} in a breeder 'blanket', which surrounds the core. After a period of time known as the **doubling time** (about 10–20 years), as much fissile material is produced in the blanket as has been consumed in the reactor. This material may then be extracted and separated for use as a nuclear fuel. In this way virtually *all* the uranium in the uranium ore may be converted to fissile material instead of less than 1%.

Full-scale fast reactors have been constructed in France, USSR and England. The disadvantage is that they are vastly more complex and expensive than burner reactors.

Fusion reactors

Fusion reactors obtain energy by virtually reproducing a sun here on earth. They are still in the experimental stage, with current research based on the fusion reaction between two isotopes[*] of hydrogen (deuterium and tritium) to form helium. The deuterium nucleus has one extra neutron and occurs in small amounts in water, from which it may be extracted and concentrated for use as a fuel. The tritium nucleus contains two extra neutrons and does not occur naturally but can be obtained from lithium isotopes in fission reactors.

In order to initiate the fusion reaction, an extremely high temperature is needed—about one hundred million degrees Celsius! At this temperature, the negatively charged electrons separate from the positively charged nucleus and form plasma. The positively charged nuclei have enough energy to overcome their tendency to repel one

---

[*] An isotope of an element has the same numberr of protons and electrons but a different number of neutrons.

another and, in collisions, they lock together to form a new nucleus; in so doing, they release nuclear fusion energy.

The technological problems involved in creating and controlling the right conditions for fusion reaction are immense and have not yet been solved.* If fusion reactors are ever brought to a successful stage of development, they hold the promise of providing all the energy needs of humans into the indefinite future. Another advantage of fusion energy is that virtually no radioactive material is formed and therefore there are no problems of disposal of nuclear waste. Also, strange as it may seem, the fusion reaction is thought to be safer than the fission reaction because in the event of a malfunction, the most likely consequence is that the plasma would be destroyed and the reaction would cease.

1.11 *TIDAL ENERGY*

Tidal energy is an interesting form of energy in that it is the only truly different form of energy available to humans other than nuclear energy and solar energy. Tidal energy results from the kinetic energy of the earth and moon, and the gravitational attraction between them and the sun. This is gradually reducing the kinetic energy of the earth and moon. The energy available from tidal flow can be used to drive turbines in much the same way as hydroelectric power generation. There are no technological problems, but some ecological changes may result from the changes in the strength or pattern of tidal flow.

Tidal power will never be able to contribute more than a few per cent of the energy needs of the world. However, power stations using tidal flow have been constructed in France, USSR, Canada and the United Kingdom but there are no plants in Australia. Several suitable sites are available, for example in the Kimberley Region of Western Australia, but due to the remoteness of the location, utilisation of this power is uneconomic.

1.12 *GEOTHERMAL ENERGY*

Geothermal energy is heat generated within the earth. Some of this heat is generated within the crust of the earth, and some is generated at greater depths and conveyed upward by movement of molten rocks. Some geothermal energy derives from the heat associated with the initial formation of the earth and some is generated continually as a result of the decay of radioactive elements within the earth's core. When geothermal energy is used for power production on an appreciable scale, the extraction rate may exceed the natural recovery rate, and hence geothermal energy cannot be regarded as a totally renewable energy resource.

Economic recovery of geothermal energy depends on the thermal gradient. Normally this is of the order of 10–40°C per kilometre of depth and is not sufficient to make recovery viable. However, gradients of 70°C or even higher exist in some areas (particularly those in the seismic belt) and recovery then becomes economic. Recovery may be based on natural steam formation (wet field) or by pumping water down to the hot depth through a bore hole and recovering the steam from a second bore hole (dry field). The steam can be used to drive a turbine and generate electricity in the conventional way.

---

* Claims to the creation of a cold fusion reaction under laboratory conditions have not been substantiated.

At the present time Australia does not use geothermal energy, but in New Zealand about 200 MW of power are currently obtained with this method. Other overseas countries such as USA, Japan and Italy generate appreciable amounts of power from geothermal energy. There are some environmental problems because chemical contaminants (hydrogen sulfide gas and carbon dioxide gas) are released in appreciable quantities from the well. Also, there can be considerable noise from the wells.

1.13 *FOSSIL FUELS*

Fossil fuels are mainly coal, oil and gas, fuels derived from the fossilised remains of plants and animals that lived on earth millions of year ago. They are non-renewable fuels because the conditions that existed during the evolution of the earth are not being repeated and because an extremely long time frame is needed for the process. Fossil fuels are relatively cheap and easy to use but their energy is recovered by the process of combustion, which may produce many pollutants that have an adverse effect on human health and the environment. Even with the most careful control over the combustion process and the application of state-of-the-art pollution-control technology, the production of carbon dioxide cannot be avoided. The popularity of fossil fuels means that carbon dioxide is produced in huge quantities and this is contributing significantly to the 'greenhouse effect' responsible for global warming. This could have serious long-term effects on the ecology of the earth.

Coal

Coal is the most abundant of all the fossil fuels and accounts for about three-quarters of the total world fossil-fuel reserve. Australia is particularly fortunate in having extensive coal fields. The majority of the coal used in Australia (and throughout the world) is for the generation of electrical power. In the past, coal was also used for heating (in homes or industry), for steam generation in boilers and as the energy source for steam locomotives. Pollution problems and the low efficiency of steam engines caused a demise in the use of coal for these purposes except in some nations such as India, where coal-fired steam locomotives are still used.

The difficulties associated with the combustion of coal at the point of usage can be overcome in several ways as now described.

Use electrical energy   This is the most common way of using the energy content of coal without the difficulties associated with the combustion of coal. Coal is used to generate electrical energy in large power stations where the combustion process and the products of combustion can be carefully controlled. The electrical energy is fed through transmission lines to the consumer, where it may be used for heating (with induction or resistance heaters), to provide mechanical energy (using an electric motor) or for other purposes that require electricity. Electricity obtained from overhead wires or rails provides the energy that drives many urban transport systems using electrified trains, trams and trolley cars. This is not an acceptable solution for motor vehicles, which do not operate on fixed routes. However, by use of storage batteries, motor vehicles may be completely mobile but the batteries need to be recharged at periodic intervals. The power–weight ratio is low and the costs are such that electrically driven motor vehicles are not very popular at present.

Convert coal to an oil or a gas   In Australia, 'town gas' (mainly methane and hydrogen) derived from coal was used as a gaseous fuel in homes and businesses for many decades before being supplanted by natural gas. There have also been proposals for the underground gasification of coal seams to tap the gas rather than mine the coal but these proposals have not yet come to commercial fruition.

Coal may also be converted into oil, and several overseas plants are in operation (for example in Germany) but the process is wasteful and yields only about 30% of the original energy content.

Pulverise coal and mix it with oil or water   Research continues into pulverised coal–oil or coal–water mixtures as a liquid fuel. This solution avoids the wastage of energy associated with coal–oil conversion but as yet can only be considered as being in the experimental stage.

Oil

Oil exists as a liquid deposit in many places in the earth's crust, where it can be recovered by drilling. It is often mixed with earth and shale, in which case the recovery process is more complex and costly. Before oil can be used as a lubricant or fuel, it needs to be refined and this is essentially a distillation process. Refining produces fractions of varying density, viscosity and energy content known collectively as petroleum. Heavy and waxy oils are used as fuel in large diesel engines, lighter fuel oils are used for smaller diesel engines, kerosene is used as an aircraft fuel and petrol as a fuel for spark-ignition engines used in many motor cars and other power plants. Liquid petroleum fuels are by far the most popular energy source for applications requiring mobility such as motor cars, trucks, ships and aircraft, and the usage is such that these fuels will be in short supply long before there is a shortage of coal.

A by-product of the refining process is a gas known as petroleum gas, which was originally simply flared off (and often still is). However, this waste can be avoided if the gas is compressed and liquefied (LPG) and this is a suitable fuel for many applications including heat engines.

Gas

Oil fields have pockets of gas above the oil—this gas is known as 'natural gas'. In the past, like petroleum gas, this was usually flared off but, nowadays, if the gas exists in appreciable amounts, it is worthwhile to recover the gas. The best method is to pipe the gas through pipelines to consumers and many such pipelines have been built in Australia and throughout the world. When pipelines are not viable, the gas may be compressed and liquefied (LNG) and then transported in pressure vessels to the point of use.

1.14 *DEPLETION OF STORED FUEL RESERVES*

By far the greatest usage of energy by people is provided by fossil or fissile fuels. These are in fixed supply and non-renewable so at some point in time they will be depleted. The question is when? One scenario is scenario A shown in Figure 1.10. This scenario assumes a fixed reserve and a usage that extrapolates from the current known rate. However, this scenario is unlikely because as supply dwindles, prices increase. This in turn tends to increase the known reserve because:

- exploration intensifies and new reserves are discovered;
- reserves with marginal viability become economic (for example shale-oil).

Also the demand decreases because:

- conservation measures are stepped-up;
- renewable energy becomes more viable.

The combination of these effects is shown in scenario B in Figure 1.10, the more likely scenario, which indicates that depletion dates will extend further into the future than are suggested by scenario A.

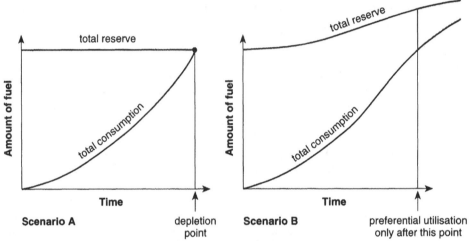

Fig. 1.10 *Two scenarios for depletion of stored fuel reserves*

Apart from energy usage, coal and petroleum provide the raw materials used in the manufacture of many products such as steel and plastics. At some future time, their use as an energy source may become restricted and they may be reserved exclusively for manufacturing materials. This situation may occur with petroleum within a relatively short space of time (less than 50 years) but coal and uranium reserves in Australia (and the world in general) should last for a least one or two more centuries.

 ## Self-test problem 1.2

(a) Outline briefly *two* disadvantages of using coal directly as a fuel in transportation.
(b) Outline briefly *two* methods by which coal can be converted to a more useful transportation fuel.
(c) When oil/gas supplies run out the following methods of providing energy needed for transportation are possible: A solar (direct), B wind, C electrical, D nuclear, E alcohol and F coal (directly used).

   A list of transportation methods is given below. Use the letters A to F to indicate which methods are suitable (using existing technology):

   motor cars (urban)
   rail
   boats (small)
   ships (large)
   aircraft (large)

1.15 *ENERGY CONSERVATION*

The increasing cost of energy provides the economic incentive to practise energy conservation. This may be done in five main ways as listed below:

- Reduce energy usage.
- Recycle materials wherever possible.
- Improve conversion efficiencies.
- Use low-grade energy currently being wasted.
- Use renewable energy wherever possible.

Reduce energy usage

When energy is cheap and plentiful, people tend to use it with little thought toward conservation. There is no doubt that energy usage can be reduced and energy wastage avoided in many ways. The starting point to an energy conservation program should be an energy audit to establish where energy is being used, when and why. From this, causes of energy waste could often be identified and usage reduced. Some methods of reducing energy usage and waste are listed below:

- Ensure lights, heaters, air conditioners and other appliances are used only when necessary and for as long as necessary.
- Ensure correct temperature control of heaters, air conditioners, refrigerators and so on. These often use more energy than is really required.
- Design buildings (or retro-fit existing buildings) to maximise the use of natural energy sources to provide comfortable temperatures. Indeed, with correct design and inclusion of appropriate insulation, shades, awnings, blinds and so on, it is possible that very little additional energy may be needed to provide comfortable year-round temperatures in many locations in Australia.
- Use rail and sea transportation and urban transport systems wherever possible in preference to motor cars and trucks.
- Reduce recreational energy usage, for example on sporting and recreational activities such as motor sports, which use large amounts of energy.
- Use smaller motor vehicles for single-person commuting.
- Use diesel rather than petrol engines and manual rather than automatic transmissions wherever possible.
- Ensure adequate insulation of all equipment that requires elevated or reduced temperatures (for example boilers and steam lines).
- Ensure fluid systems are designed to minimise heat loss by use of smooth, large-diameter pipes and large-radius fittings.
- Use compact fluorescent lights in place of incandescent lights.
- Use power-factor-correction equipment in industrial applications that use large amounts of electricity.
- Replace old, inefficient equipment with up-to-date fuel-efficient plant.

Recycle materials wherever possible

Many materials can be recycled; this requires much less energy than that required for the extraction and processing of raw material. For example, about 90 MJ of energy is necessary to manufacture 1 kg of aluminium from ore, whereas only about 1 MJ is necessary to recycle aluminium. The feasibility of recycling metals, glass, paper products and some plastics is established. The main problem is that of collection and separation, and this in turn depends on attitudes and co-operation as well as economic incentives. Another approach, which does not depend on human co-operation, is the use of mechanical or chemical separation processes for domestic and industrial waste. This is feasible but is seldom economically viable (at present).

Improve conversion efficiencies

Great strides have already been made by engineers in improving the efficiency of energy-conversion devices. For example, the efficiency of heat engines and power-generation equipment has been improved dramatically in the past 50 years or so. The designs of motor vehicles, aircraft and ships have all improved, resulting in the reduction of friction and drag and an increase in efficiency.

Research continues, and no doubt efficiencies will continue to improve. For example, if the development of an internal-combustion engine using ceramic components were successful, the efficiency of these engines would be considerably improved. The MHD (magnetohydrodynamic) process currently undergoing research in Australia and overseas involves electricity generation from a hot stream of ionised gas without the use of a heat engine. This process offers the prospect of electricity generation with efficiencies 10–15% higher than those currently being achieved.

Despite the achievements and the prospect of new processes offering even higher efficiencies, there is no doubt that the law of diminishing returns applies, and efficiency gains are increasingly more difficult to obtain. In some cases, efficiency has been pushed to such a high level that further gain can only be minuscule, as with water turbines and steam boilers. If a heat engine is involved, then the energy-conversion efficiency suffers the limitation of the maximum possible efficiency (Carnot efficiency), which depends on temperature and is considerably less than 100%. Since this is a fundamental law of nature (the second law of thermodynamics) it cannot be overcome.

Use low-grade energy currently being wasted

When a heat engine (of any type) is used, less than half the energy of the fuel is converted to useful work and more than half is lost. If only a proportion of this lost energy can be recovered, vast energy savings are possible.

One method, becoming increasingly common, is to use hot exhaust gases to power turbochargers, which compress the incoming air to the engine and enable higher efficiencies and power outputs to be obtained. Another method, known as **cogeneration**, is the combined on-site generation of mechanical energy and useful heat energy. In cogeneration, the energy normally lost from a heat engine is put to a useful purpose and this reduces the overall on-site energy consumption. Cogeneration has been a common practice in ships for many years. If the ship is powered by a diesel engine, after turbocharging the remaining energy in the exhaust gas is used to power auxilliaries (such as electric generators) and to provide hot water for the ship. In this way, most of the energy in the fuel is effectively utilised. Cogeneration is becoming increasingly common in industrial plants, where a heat engine is used with a generator to provide the electrical energy needs. The hot exhaust gases are used to generate steam or to provide the heat needed for industrial processes in the plant. The heat may even be used for cooling purposes by use of an absorption refrigeration plant.

Large power stations in Australia are usually located at some distance from high-energy consumers, which makes it uneconomic to use hot water from the power stations for any useful purpose. However, in some overseas countries such as Scandinavia, distances are smaller and hot water from power stations is piped to cities where it is used for home heating and even for de-icing roads in winter.

Use renewable energy wherever possible

Most of the renewable energy sources discussed in this chapter have proven feasibility. The main difficulties in their use are cost, convenience and availability. There is no doubt that price escalation and decreasing availability of non-renewable energy reserves will mean increasing use of renewable energy and this will help to conserve existing non-renewable reserves.

Summary

Energy is essential for life on earth. The use of human (or animal) muscle power to do work is of strictly limited scope and in order to raise productivity and standard of living, other energy sources need to be employed. These fall into two main classes:

- Renewable energy, which is essentially solar energy in one of its many forms.

- Non-renewable energy, which may be in the form of fossil fuels (coal, oil and gas) or nuclear energy.

Most energy sources provide heat, so heat energy is of special significance to humans. Apart from heat itself, the most needed form of energy is mechanical energy. The conversion of heat energy into mechanical energy requires a heat engine and the conversion efficiency is limited by the **second law of thermodynamics**, which is a natural law and cannot be overcome by any means. The energy needed to power the Industrial Revolution was derived essentially from fossil fuels and their use has escalated since then. However, they are in fixed supply and reserves are being rapidly depleted. Future generations will look on this period of time as the 'fossil-fuel age', an extinct way of life that can never be repeated on earth. An alternate source of energy is nuclear-fission energy but disposal of nuclear wastes presents an environmental nightmare. However, nuclear energy will be able to provide much of the energy needed by humans for the next several centuries. There has been much research into nuclear-fusion energy, which holds the promise of almost unlimited energy for the future, but this must be regarded as a highly speculative possibility. Renewable energy is the most acceptable from an environmental view but is usually too costly and of too low an energy density to have a significant impact on total energy requirements (at the current rate of usage).

The usage of energy can be reduced by practising energy conservation and this concept is gaining acceptance as people become more aware of the need to conserve energy and protect the environment. It is probable that future energy needs of humans on earth will not be provided by a single source but by 'niche usage', that is, by using many different sources wherever the use of a source is feasible, economic and environmentally acceptable.

Problems

1.1 State a mechanism by which the following energy conversion process can occur directly:

(a) heat → mechanical
(b) electrical → mechanical
(c) mechanical → heat
(d) mechanical → electrical
(e) solar radiation → heat
(f) solar radiation → chemical
(g) solar radiation → electrical
(h) solar radiation → mechanical
(i) heat → electrical
(j) chemical → electrical
(k) chemical → heat
(l) nuclear → heat
(m) electrical → heat

1.2 (a) Explain what is meant by the *availability of heat energy*.
(b) How can availability of heat energy be measured?
(c) How does availability of heat energy relate to temperature?
(d) How does the efficiency with which heat energy can be converted to work relate to temperature?
(e) Why is it wasteful to use combustion or electrical-resistance heating for human-comfort heating? What is being wasted in this case?

1.3 (a) What initial energy source has provided all energy available for humans?
(b) State *six* different forms of solar energy and describe briefly a method by which each form may be used.
(c) Does tidal energy result from solar radiation? Explain.
(d) State *two* energy forms that can be stored for an indefinite time without appreciable loss.

1.4 (a) Explain what is meant by *photosynthesis*.
(b) State *two* advantages and *two* disadvantages of using photosynthesis to provide energy.
(c) What energy is needed for photosynthesis, and into what form is this energy converted?
(d) State *three* ways that the energy obtained by photosynthesis may be utilised by humans.

1.5 (a) State typical values of radiant solar energy in kW/m^2 received in Australia:
(i) outside the atmosphere;
(ii) at the surface of the earth on a clear day.
Also give the *average* value of radiant solar energy received during daylight hours over several months.
(b) State *two* advantages and *two* disadvantages of direct solar energy as an energy source.
(c) What is the function of a solar cell, and what are its main advantages and disadvantages?
(d) What are *two* ways by which solar heat could be used for cooling applications?

1.6 (a) Draw neat sketches of cross-sections through both a typical flat-plate solar collector and a focusing collector.

(b) What is the main advantage and the main disadvantage of a focusing collector compared with a flat-plate collector?

1.7 (a) With the aid of a neat sketch, describe the operation of a power tower. Also state its main advantages and disadvantages.

(b) With the aid of a neat sketch, describe the operation of the ocean thermal-energy collector. Also state its main advantages and disadvantages.

1.8 (a) Explain the operation of the solar pond, and state its main advantages and disadvantages.

(b) What is the source of geothermal energy, and how can this energy be used to generate electricity?

1.9 (a) With the aid of a neat sketch, explain the operation of a fission-reactor power plant of the conventional type.

(b) How does a *breeder* reactor differ from a *burner* reactor, and what are its relative advantages and disadvantages?

1.10 (a) Transmission lines are not viable for transmitting electricity over long distances. State an alternative method that overcomes the disadvantage of distance and outline the energy-conversion steps involved.

(b) Explain the energy-conversion steps involved in hydroelectric power generation. State *two* advantages and disadvantages of this method of power generation compared with a coal-fired power station.

(c) What is the energy source for tidal energy, and how can this energy be used to generate electricity?

1.11 (a) State *two* reasons why sailing ships are no longer used for commercial transportation purposes. Explain how wind energy can be used for this purpose without incurring these disadvantages.

(b) Explain *two* methods by which wave energy can be converted to electrical energy. State *two* advantages and disadvantages of the use of wave energy for power generation compared with a coal-fired power station.

1.12 (a) Explain the fusion process for obtaining nuclear energy. What are its advantages and limitations when compared with the fission process?

(b) At the present time, there are many large-scale nuclear power plants operating throughout the world but none in Australia. Explain the main reasons for this.

1.13 (a) Describe briefly the source and formation of fossil fuels.

(b) Describe briefly the *three* main types of fossil fuels and their main uses.

(c) How has the energy of fossil fuels been utilised, and what has been the ecological impact of this method?

1.14 (a) With the aid of a neat graph, plot *total fossil-fuel reserves* against *time*, showing:

(i) the scenario where supply remains fixed and demand continues in the same way;

(ii) the more likely scenario.

(b) Explain the *two* main reasons why scenario (i) is unlikely to eventuate.

1.15 State *five* methods of conserving non-renewable energy reserves. Under each heading, give at least *three* examples to illustrate your answer.

1.16 Describe briefly at least *six* different ways by which energy usage and waste could be reduced using existing technology.

1.17 **(a)** Explain in terms of energy utilisation why it is unlikely that we will return to riding horses when fossil fuels are in short supply.

(b) Describe briefly four alternatives for human transportation which do not require energy from fossil fuels and which do not involve unproven technology. State at least one significant advantage and one significant disadvantage for each alternative you propose.

1.18 Imagine the world in the year 2050. Petroleum fuels (oil and gas) have been depleted to the point where they can no longer be used as a source of energy. Attempts to use the nuclear-fusion process have not been successful and no other energy source not known today has been discovered.

Describe briefly at least two feasible energy sources and conversion processes which could be used in Australia for the five activities listed below:

(a) home lighting and cooking;

(b) maintaining a comfortable temperature in buildings and homes;

(c) overseas transportation (sea and air);

(d) overland transportation;

(e) short-distance transportation (commuting, shopping, etc.).

1.19 Extend the scenario given in Problem 1.18 to the year 2200 when coal stocks have also been depleted. All other factors remain the same.

Basic concepts

Introduction

In this chapter the basic concepts necessary for a study of thermodynamics are treated. Some of these concepts, such as those of a system or a process, may be described or defined but cannot be measured. Other concepts, such as pressure or mass, may be defined and measured, and this requires a consistent system of units. In Australia, the system of units that has been adopted is the SI system and this system is used throughout the book. It is most important that correct units are maintained in all equation substitutions and calculations. Indeed, a significant part of problem solving in thermodynamics is meticulous use of the correct units.

Notes

1 For greatest accuracy in calculation, wherever possible in this book, values are retained in calculator memory during calculation steps. Answers are given in bold and are rounded off to three (or sometimes four) significant figures.

2 The shorthand method $A(d)$ is used in calculations to indicate the area of a circle. For example, the area of a 0.3 m diameter circle will be shown as $A(0.3)$ rather than as

$$\pi \times \frac{0.3^2}{4}.$$

2.1 *THE NATURE OF MATTER*

To the human eye, matter appears homogeneous; for example, a block of metal or a drop of water looks like a continuous chunk of matter. However, all matter is in fact composed of a huge number of discrete particles known as **molecules**. The molecules are themselves composed of simpler elementary particles called **atoms**. The composition of the atoms in the molecule is expressed by a chemical formula. Water, with the chemical formula H_2O, is composed of two atoms of hydrogen and one of oxygen; carbon dioxide (CO_2) is composed of one atom of carbon and two atoms of oxygen. However, the chemical formula gives no indication of how the atoms are structured in the molecule; for example, in carbon dioxide the two atoms of oxygen lie opposite one another, whereas in water they are inclined to the hydrogen atom at an angle of 104° (Fig. 2.1).

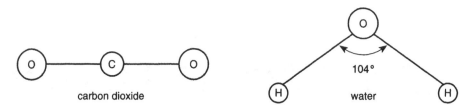

Fig. 2.1   *Molecular structure of carbon dioxide and water*

The mass of a molecule is measured by the **relative molecular mass** (formerly known as molecular weight). This is the mass of the molecule *relative* to the mass of a natural carbon atom taken as exactly 12. *Precise* measurement of relative molecular mass of other elements shows small variations from simple whole numbers due to the presence of naturally occurring isotopes and the fact that the mass of a proton or neutron varies slightly from one element to another. However, for all purposes except calculations involving nuclear reactions, simple whole numbers as given in Appendix 3 may be used, resulting in negligible error.

Since a substance appears homogeneous and the molecules cannot be seen, there must be a very large number of them in even a small amount of matter. This is so; in fact there are about 33.5×10^{24} molecules in 1 L (or 1 kg) of water. To gain a better concept of this huge number, imagine a molecule-counting machine that could count one million molecules per second. Suppose the machine counted the number of molecules in 1 L of water. How long would it take before the count was finished? 1.06×10^{12} years, or over one million million years!

States of matter

Matter may exist in any one of three states, namely as a solid, liquid or gas. These states are also known as **phases** and a change in state is also known as a phase change. Liquids and gases are also known as **fluids**.

The states of matter in terms of the molecular model are illustrated in Figure 2.2.

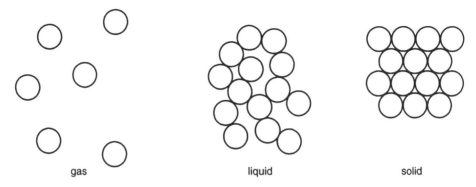

Fig. 2.2 *Simple molecular model of the three possible states of a substance*

Solids

Solids have rigidity because the molecules are close together and interlocked so that they remain in the same relative positions to one another.

Gases

In gases the molecules are far apart and the forces between them are negligible so that the molecules move independently of one another (except in a collision).

Liquids

Liquids are intermediate between solids and gases, the molecules are free to move and are closer together than in a gas yet they are not tightly packed as in a solid. The molecules are said to be cohesive (stick together) but are still capable of motion relative to each other. This enables liquids to flow because when a liquid flows, the molecules slide over each other.

Molecular motion

The molecules of all substances (at a temperature above absolute zero) are not still but are in continual motion.

In a gas the molecules are capable of independent motion in several modes, namely **translational**, **rotational** and **vibrational** (Fig 2.3). These modes can take place simultaneously and also in three dimensions.

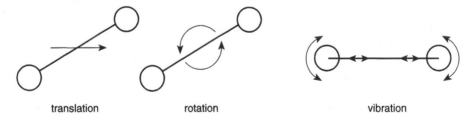

Fig. 2.3 *Modes of molecular motion*

In a solid, the molecules can rotate and vibrate but there is no translation, so the molecules still retain the same spatial orientation relative to one another. In a liquid, translation occurs, and the molecules may move relative to one another but they do not move away from each other, that is, they remain cohesive.

Phase change

Most substances can exist in each of the three states, namely as a solid, a liquid or a gas. A phase change from one state to another occurs as a result of energy transfer into or out of the substance. This is illustrated in Figure 2.4.

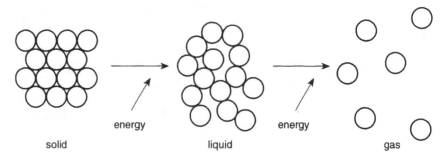

solid   energy   liquid   energy   gas

Fig. 2.4 *Energy is needed for a phase change*

The energy possessed by the molecules of a substance is known as their **internal energy**. Gases have the highest internal energy and solids the least, with liquids intermediate. As energy (in the form of heat) is input, the molecules of a solid become more energetic and are eventually able to break the bonds that bind them firmly together. When this occurs, the solid starts to become a liquid. Further energy input enables the molecules to break completely free of each other and to form a gas.

The path shown in Figure 2.4 may be reversed if energy is transferred out of the substance so that the internal energy of the molecules decreases.

2.2 PROPERTIES AND PROCESSES

The properties of a substance are the quantities that may be measured or calculated and give meaningful information about the state of the substance but are independent of the path (or method) taken to achieve their value. For example, temperature is a property because if the temperature of a block of steel is 50°C, this gives meaningful information by itself. It is not necessary to know if the block reached this temperature by being heated from a lower temperature or cooled from a higher temperature, or how the heating or cooling was done. Indeed the block may have reached 50°C as a result of being machined. Whatever the process or path taken, this has no effect on the property.

The path taken to change a property is known as a **process**. In thermodynamics, processes usually involve the transfer of energy such as heating or cooling, compressing or expanding, stirring or pumping. The amount of energy transferred depends on the process as well as the end states (properties). For example, it is possible to transfer different amounts of energy by different processes and still maintain the same initial and final end states (properties).

It is now appropriate to look at common thermodynamic properties of a substance, such as mass, volume, density and so on.

2.3 MASS

Mass (m) is the quantity of matter in a substance. Because substances are composed of molecules, the total mass is the sum of the individual masses of all the molecules. The unit of mass is the kilogram (kg) or tonne (t); 1 t = 1000 kg. Mass does not change with position, nor is it affected by pressure, temperature or motion (unless the velocity approaches the speed of light—Einstein's relativity theory). If a solid is heated until it burns, the mass *appears* to change, as the pile of ash left is much less than the original substance. What is happening here is that much of the solid matter is converted to gas, but the *total* mass does not change. Indeed, one of the most fundamental laws of nature is that *mass cannot be created or destroyed* and this is known as the **law of conservation of mass**.

As a result of Einstein's work, it is now known that mass and energy are mutually convertible, so this law should more correctly be called the law of conservation of mass and energy. However, the mass change is extremely small in engineering work and there is insignificant error in using these laws separately. So throughout this book, conservation of mass and conservation of energy will be used as separate laws (except in problems involving nuclear reactions).

2.4 VOLUME

The volume (V) of a body is the amount of space it occupies. Since the unit of length in the SI system is the metre (m), the unit of volume is the cubic metre (m^3). For many practical purposes, m^3 is too large a unit so the litre (L) is often used (although not strictly an SI unit).

The following relationships are often useful:

$$1000 \text{ cubic millimetres } (mm^3) = 1 \text{ millilitre } (mL)$$
$$1000 \text{ millilitres } (mL) = 1 \text{ litre } (L)$$
$$1000 \text{ litres } (L) = 1 \text{ cubic metre } (m^3)$$
$$\therefore 10^9 \text{ mm}^3 = 1 \text{ m}^3$$

Unlike mass, volume may be altered by changes in pressure and temperature (particularly in the case of gases).

2.5 DENSITY

Density (ρ) is the mass per unit volume.

$$\rho = \frac{m}{V}$$

density (2.1)

SI units of density are kg/m^3.

As the name suggests, density is a measure of the size of the molecules and how closely the molecules are spaced in a material. For example, when we say lead is heavier than aluminium, we really mean that lead has a higher density than aluminium. Note that the density of water (at standard temperature and pressure)* is $1000 \text{ kg/m}^3 = 1 \text{ kg/L}$.

---

* Standard temperature is 0°C and standard pressure is 101.325 kPa.

2.6 *RELATIVE DENSITY*

Relative density (RD) is the ratio of the density of a substance to the density of water (at standard conditions).

$$RD \text{ (substance)} = \frac{\rho \text{ (substance)}}{\rho \text{ (water)}}$$

relative density (2.2)

Because relative density is a ratio, it has no units.

Example 2.1

The *RD* of mercury is 13.6. Calculate its density.

Solution

$$\rho \text{ (mercury)} = RD \text{ (mercury)} \times \rho \text{ (water)}$$
$$= 13.6 \times 1000$$
$$= \mathbf{13.6 \times 10^3 \ kg/m^3}$$

2.7 *SPECIFIC VOLUME*

In thermodynamics, 'specific' usually means *per unit mass*. Hence *specific volume (v) is the volume per unit mass.*

$$v = \frac{V}{m} = \frac{1}{\rho}$$

specific volume (2.3)

The units of specific volume are cubic metres per kilogram (m^3/kg). Specific volume is used mainly for gases.

2.8 *FORCE*

Force (F) is 'push' or 'pull', or is often loosely described as 'tension' or 'pressure'. Newton realised that force is always associated with a tendency for motion or, more particularly, an acceleration. In fact his famous equation

$$F = ma$$

force (2.4)

means that force is not a basic unit since it is dependent on the units for mass and acceleration. In the SI system, the unit of force is kilogram metres per second squared (kg m/s^2). Since this is awkward, it is called a **newton** (symbol N).

One newton is a small force (about the weight of a small apple), so for many practical problems, kilonewtons (1 kN = 10^3 N) are used.

2.9 *WEIGHT*

Weight (w_t) is the gravitational force on a body.

$$\boxed{w_t = mg}$$

weight (2.5)

where g is the gravitational constant or acceleration due to gravity (9.81 N/kg or m/s^2) on the earth's surface. Although g and therefore w_t vary with location and height on the earth, the standard value quoted for g will suffice for most problems encountered.

Since weight is a force, units of weight are newtons.

2.10 *PRESSURE*

Pressure (p) is the force per unit area.

$$\boxed{p = \frac{F}{A}}$$

pressure (2.6)

The units are newtons per square metre (N/m^2), called **pascals** (Pa) in the SI system. One newton is a small force and 1 m^2 is a large area; therefore 1 Pa is a very small pressure and so pressures are usually given in kilopascals (1 kPa = 10^3 Pa) or MPa (1 MPa = 10^6 Pa).[*]

Atmospheric pressure

Atmospheric pressure (p_{atm}) is the pressure associated with the atmosphere due to the weight of air. Although this pressure varies according to location and weather patterns, an average value at sea-level is 101.3 kPa. This is also called 'normal' atmospheric pressure.

Gauge pressure

Gauge pressure (p_g) is the amount by which pressure differs from atmospheric pressure. This is measured with a gauge that measures the pressure above (or below) atmospheric pressure. The gauge pressure below atmospheric is negative and called **vacuum**.

Absolute pressure

Although there in no limit to how high a pressure can be, there is a limit to how low it can be. This point of absolute minimum is the **absolute zero of pressure** (no pressure at all). *Absolute pressure (p) is pressure measured above this zero point.* Therefore

$$\boxed{p = p_g + p_{atm}}$$

absolute pressure (2.7)

Note When using equations or solving problems, care needs to be taken to distinguish between gauge pressure and absolute pressure. Thermodynamic equations usually use absolute pressure whereas fluid mechanics equations usually use gauge pressure. Conversion from gauge to absolute or the reverse may be necessary before substitution in an equation that involves pressure.

---

[*] 1 MPa is the pressure due to a weight of 1 N on an area of 1 mm^2, that is, the pressure due to an apple (1 N) resting on a vertical matchstick (1 mm × 1 mm).

Relationship between pressure and height

For a substance of constant density (such as a liquid), the pressure at any vertical position due to the self-weight of the substance above that datum is independent of the surface area and is given by the equation:

$$p = \rho g h$$

pressure increase due to self-weight of a liquid (2.8)

where p = pressure (Pa)
ρ = density of the liquid (kg/m^3)
g = gravitational constant (acceleration due to gravity) (N/kg or m/s^2)
h = depth below the free surface

Notes

1   This equation applies to substances in which there are negligible density changes (incompressible). It applies to liquids in any shape of container and also to solids that have a uniform vertical cross-section throughout.

2   The pressure given by this equation is the *increase* in pressure due to self-weight. If the free surface is at atmospheric pressure and this is taken to be zero (gauge), then this equation will give the *gauge* pressure at depth h. If the free surface is at any pressure other than zero, the total pressure at depth h will be the sum of the pressures at the surface and the pressure increase due to self-weight at depth h.

3   Proof of this equation is given in Appendix 5.

Example 2.2

A barometer reads 760 mm of mercury. Convert this to pascals.

Solution

Now ρ (mercury) = 13.6 × 1000 kg/m^3 (see Example 2.1).

$$p = \rho g h$$

$$\therefore p = 13.6 \times 1000 \times 9.81 \times \frac{760}{1000} \text{ Pa}$$

$$= 101.4 \times 10^3 \text{ Pa}$$

$$= \mathbf{101.4 \ kPa}$$

Thus 'normal' pressure (101.3 kPa) can also be expressed (very closely) as 760 mm of mercury.

Example 2.3

A water manometer is connected to the inlet manifold of an engine and records a height of 300 mm as shown (Fig 2.5). Convert this to:
(a) gauge pressure in kilopascals;
(b) absolute pressure in kilopascals;
if barometric pressure is 100.9 kPa.

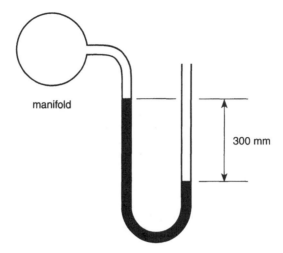

manifold

300 mm

Fig. 2.5

Solution

From the fluid height in the limbs (Fig. 2.5), it is clear that the manifold pressure is less than atmospheric pressure (vacuum).

(a)
$$p = \rho g h$$

$$\text{Gauge vacuum} = 1000 \times 9.81 \times \frac{300}{1000}$$

$$= 2943 \text{ Pa}$$

$$= 2.943 \text{ kPa}$$

or

gauge pressure = **−2.94 kPa** (rounded to three significant figures)

(b) Absolute pressure, $p = p_g + p_{atm}$
$$= -2.943 + 100.9$$
$$= 97.957$$
$$= \textbf{98 kPa} \text{ (rounded to three significant figures)}$$

Self-test problem 2.1

A steel piston of dimensions shown in Figure 2.6 (on page 38) rests above gas in a cylinder. The relative density of steel may be taken as 7.8 and friction between the piston and the cylinder may be neglected.

Determine:

(a) the gauge pressure of the gas in the cylinder—hint: this is the same as the pressure at the base of the piston due to its self-weight;

(b) the absolute pressure of the gas in the cylinder if atmospheric pressure is normal.

(c) Recalculate the gauge pressure and absolute pressure of the gas if a mass of 5 kg is now placed on top of the piston.

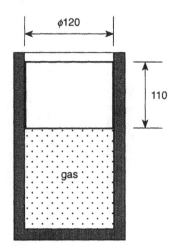

Fig. 2.6

2.11 *TEMPERATURE*

Temperature is a difficult concept to define but because the human body senses temperature and responds to it, we all have an intuitive feeling for temperature expressed as 'hotness' or 'coldness'.

At the microscopic level, *temperature (T) is a measure of the average linear kinetic energy of the molecules of a substance*, that is, the total linear kinetic energy of *all* the molecules divided by the number of molecules. Since kinetic energy is energy due to motion, it is evident that the motion of the molecules will increase with increasing temperature. At absolute zero there will be no motion, and the molecules will be completely still.

Temperature may be measured using a standard scale, namely the Celsius scale (°C) (previously known as the Centigrade scale). The Celsius scale takes the boiling point of water (at standard atmospheric pressure) as 100°C and the freezing point as 0°C.

Absolute temperature

Just as pressure has no upper limit but has a definite lower limit, so too with temperature. This lower limit is called **absolute zero** and corresponds to –273°C. (A more exact value is –273.15°C, but for most purposes, little accuracy is lost by using –273°C.)

The unit of absolute temperature in the SI system is the kelvin (K). The spacing of temperatures is the same in the Kelvin scale as in the Celsius scale, but the zero point is different. A comparison of the scales is shown:

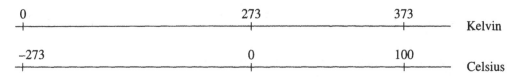

Conversion is by Equation 2.9:

$$T(\text{K}) = 273 + T(°\text{C})$$

absolute temperature (2.9)

Notes

1 Thermodynamic equations usually use absolute temperatures and, when they do, conversion from Celsius temperature is essential. The exception to this rule is for a temperature difference $(T_2 - T_1)$. Here, conversion to K is unnecessary since 273 is added to both T_2 and T_1 and cancels out in $(T_2 - T_1)$.

2 Standard temperature is 0°C or 273 K.

3 Normal temperature (temperature of an average day) is usually taken to be 20°C in Australia.

Self-test problem 2.2

(a) Convert the following temperatures to absolute:
  −269°C, 15°C, 224°C

(b) Convert the following temperatures to °C:
  15 K, 226 K, 527 K

(c) Determine the temperature change $(T_2 - T_1)$ in each of the following cases:

  initial temperature 15°C   final temperature −25°C
  initial temperature −5°C   final temperature 35°C
  initial temperature −80°C   final temperature −20°C

2.12 *SYSTEM AND BLACK-BOX ANALYSIS OF A SYSTEM*

In thermodynamics (as with most engineering and science) it is necessary to theoretically isolate some components or machines from their surroundings. For example, in order to analyse the energy transfers in a heat engine, it is necessary to treat the engine as a distinct unit. This may be done by placing a boundary around the engine, and the system being considered (the engine) is contained within the boundary. *A system may be defined as any region of space or any quantity of matter that has been separated from its surroundings.* Alternatively, a system may be thought of as an interacting or interdependent group of components that forms a unified whole.

For a heat engine, it is logical to use the physical boundaries of the engine itself to define the system. However, it is also possible to use a conceptual boundary located in space around the engine. The boundary should not be selected to pass somewhere through the engine itself as this would alter the fundamental organisation of the components (the engine would not function if so cut).

As another example of a system, consider boiling water in an electric kettle as shown in Figure 2.7 (page 40). The boundary is indicated by the broken line and the energy and mass transfers are shown.

The boundary does not necessarily have to be rigid, and by using a flexible boundary as shown in Figure 2.8 (on page 40), the mass of the system remains constant. By careful choice of boundary, analysis may be simplified.

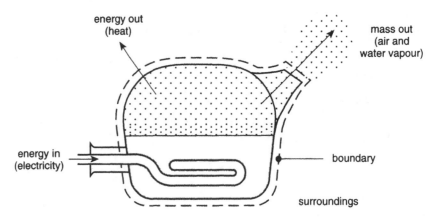

Fig. 2.7 *System for an electric kettle*

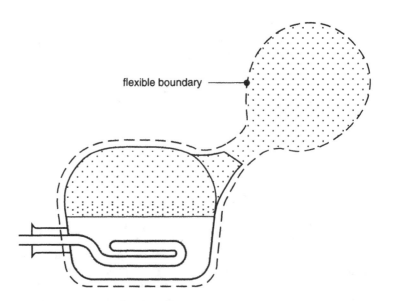

Fig. 2.8 *An alternative boundary location for an electric kettle*

At this stage the following question might well be asked: Is there some loss of accuracy or validity with the systems approach because of the introduction of an artificial boundary that may not really exist? The answer is 'no', provided that the boundary is chosen so as not to alter the fundamental organisation of the components, and that all transfers of matter and energy across the boundary are considered.

Once the system is defined, it may be analysed. One method is called **black-box analysis** because (as the name implies) internal workings of the system are not known and not investigated. Instead, the *external* relationships between inputs and outputs is examined. A black-box analysis of a heat engine is shown in Figure 2.9.

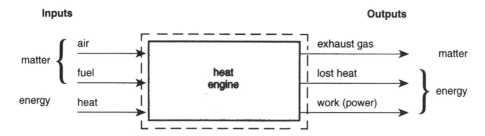

Fig. 2.9 *Black-box analysis of a heat engine*

Although the black-box method is very useful from a user point of view, it can only provide limited information when analysing the system. Often it is necessary to look 'inside the box' to obtain a more complete picture of the relationship between the variables involved.

In this book, both approaches are used. Initially, black-box analysis will be used to establish the fundamental relationships between matter and energy transfers in systems such as heat exchangers and heat engines. This will then be followed by a more detailed look inside the system to establish useful relationships between the important properties.

2.13 *RECIPROCATING PISTON-AND-CYLINDER MECHANISM*

The reciprocating piston-and-cylinder mechanism is illustrated in Figure 2.10 (on page 42), and is widely used in heat engines, pumps, compressors and fluid motors. As shown, it may be either single acting or double acting. The double-acting mechanism is more compact but has numerous disadvantages and so it is far less common than the single-acting mechanism. Therefore, single-acting operation should always be assumed unless stated otherwise.

The mechanism can be made two-stroke or four-stroke depending on the number of strokes of the piston to complete the cycle. Two strokes of the piston occur with each revolution of the crankshaft. There may be one or more cylinders out of phase with one another and arranged in many different ways, such as in-line, opposed or vee. The greater the number of cylinders, the smoother the torque fluctuations on the crankshaft and the smaller the flywheel necessary.

Terms used with the mechanism are now described.

TDC

TDC is the 'top dead centre' position or top limit of travel for the piston. It is also known as the 'outer dead centre'. At this position, the piston instantaneously stops and reverses its direction of travel.

BDC

BBC is the 'bottom dead centre' position or bottom limit of travel for the piston. It is also known as the 'inner dead centre'. Again, the piston instantaneously stops and reverses its direction of travel.

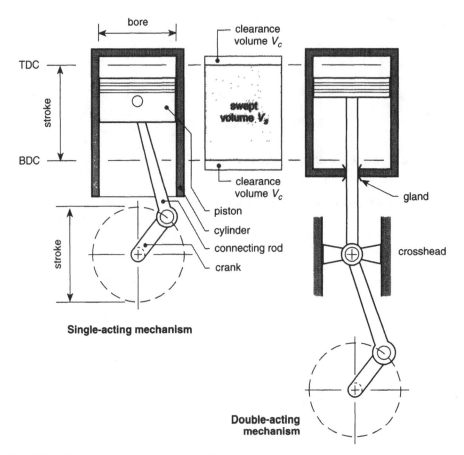

Fig. 2.10   *Reciprocating piston-and-cylinder mechanism*

Bore

The bore is the cylinder diameter. In practice, there is a small clearance between the piston and cylinder but this is usually neglected and the cylinder and piston diameter are taken to be the same. The clearance may also cause some gas leakage (particularly after wear occurs) but this is minimised by the use of piston rings and may also usually be neglected.

Stroke

The stroke is the distance between the face of the piston at the TDC and BDC positions. It is also equal to the diameter of the circle swept out by the movement of the centre of the crankpin.

Swept volume

The swept volume (V_s) is the area of the bore × stroke. For a single cylinder mechanism this is also the 'capacity'. For a multicylinder mechanism, the capacity is the swept volume of each cylinder × number of cylinders. For the double-acting mechanism, the capacity of the cylinder is almost double that of the single-acting mechanism. (It is not exactly double because the capacity of the bottom side of the piston is reduced by the volume of the rod.)

Clearance volume

The clearance volume (V_c) is the volume enclosed by the face of the piston at the TDC position. For a double-acting mechanism there is clearance volume at the BDC position also, which, like the swept volume, is reduced by the volume of the rod.

Compression ratio

Compression ratio is the ratio of the volume of the cylinder contents before compression to the volume after compression (clearance volume). Assuming compression starts at the bottom of the stroke, the compression ratio is given by:

$$\text{Compression ratio} = \frac{V_1}{V_2} = \frac{V_s + V_c}{V_c}$$

compression ratio (2.10)

Because compression ratio is a ratio, it has no units.

Note The compression ratio given by equation 2.10 is actually a *theoretical* one because, in practice, inertial effects in both the valves and the air itself mean that compression may not start until the piston has moved away from the BDC position. Hence the actual compression ratio may be somewhat less than that given by Equation 2.10.

Pressure ratio

Pressure ratio is the ratio of the absolute pressure of the gas after compression to the absolute pressure before compression. Pressure ratio is generally greater than compression ratio.

$$\text{Pressure ratio} = \frac{p_2}{p_1}$$

pressure ratio (2.11)

Because pressure ratio is a ratio, it has no units.

Self-test problem 2.3

A six-cylinder engine has a bore of 90 mm and a stroke of 100 mm, a compression ratio of 9.5:1 and runs at 5000 rpm. Determine:
(a) the capacity of the engine in litres;
(b) the clearance volume per cylinder in litres;
(c) the number of strokes made by each piston per second.

Summary

All matter is composed of extremely small particles called molecules, which are in a continual state of motion. The mobility of this motion determines the state or phase of the substance, gas molecules being the most mobile and the molecules of a solid being the least mobile. The amount of motion and therefore the internal energy of the molecules depend on the temperature; the higher the temperature, the greater the motion and the amount of internal energy. Indeed, temperature itself is a measure of the *average* internal kinetic energy of the molecules.

In thermodynamic equations it is essential to use the correct units when substituting. Particular care should be taken with gauge and absolute pressures and temperatures. Usually in thermodynamics absolute pressures and temperatures are used.

The complexity of the real world can be simplified for analysis purposes by introducing the idea of a system, that is, by separating some part from its surroundings. Analysis of the flows of matter or energy into or out of a system without consideration of the internal mechanism in the system is known as black-box analysis and is a very useful method of initial analysis of a system.

The piston-and-cylinder mechanism is widely used and the terminology associated with this mechanism should be familiar.

 # Problems

2.1 (a) In terms of the molecular structure of matter, explain the solid, liquid and gas phases of a substance.
   (b) Discuss the *three* molecular-motion modes and include a neat sketch.
   (c) Explain the concept of *temperature* in terms of molecular motion.

2.2 (a) Explain what is meant by *system, surroundings* and *boundary.*
   (b) What is *black-box system analysis?*
   (c) Why is it necessary to use systems in thermodynamic analysis?

2.3 (a) Explain what is meant by the *properties* of a substance.
   (b) State *three* properties.
   (c) In thermodynamics, what is meant by a *process?*
   (d) State *three* processes.

2.4 (a) State the *law of conservation of mass* as applicable to general engineering.
   (b) Explain why it is necessary to modify this law to include nuclear reactions.
   (c) Restate the law so that it is valid in all cases.

2.5 (a) What does the adjective *specific* usually mean in thermodynamics?
   (b) Define *specific volume* and state its units.
   (c) Define *density* and state its units.
   (d) What is the relationship between *density* and *specific volume?*
   (e) For what type of substance is specific volume, rather than density, usually used?
   (f) Define *relative density* and state its units.

2.6 (a) Define *pressure* and state its units.
   (b) What is the difference between *absolute pressure* and *gauge pressure?*
   (c) Explain *fluid* pressure in terms of molecular motion.

2.7 (a) Draw a neat sketch of the reciprocating piston-and-cylinder mechanism (single acting).
   (b) Explain the meaning of, and mark the sketch to show, the following:
      (i) top dead centre
     (ii) bottom dead centre
    (iii) bore
     (iv) stroke
      (v) swept volume
     (vi) clearance volume

(c) Define *pressure ratio* and *compression ratio*.

(d) Sketch the double-acting mechanism, and explain its purpose and how it differs from the single-acting mechanism.

2.8 Convert the following temperatures to absolute:

(a) 100°C   (b) –50°C   (c) 220°C

Convert the following gauge pressures to absolute:

(d) –76.4 kPa   (e) 2.52 MPa   (f) 62.7 Pa

(a) 373 K   (b) 223 K   (c) 493 K   (d) 24.9 kPa   (e) 2.62 MPa   (f) 101.4 kPa

2.9 A tank 500 mm in diameter and 1200 mm long weighs 1 kN when evacuated. When filled with fluid, the weight is 1.1 kN. Calculate the density and specific volume of the fluid.

43.3 kg/m³; 0.023 m³/kg

2.10 A 100 mm cube of a substance weighs 8 N. What is its density and relative density?

815.5 kg/m³; 0.8155

2.11 A free steel piston of diameter 200 mm and length 300 mm is a sliding fit in a vertical cylinder containing a gas. If atmospheric pressure is normal, determine the gas pressure when:

(a) the piston is above the gas in the cylinder;

(b) the piston and cylinder are inverted so that the gas is above the piston.

Take the relative density of steel to be 7.7 and neglect friction.

(a) 124 kPa   (b) 78.6 kPa

2.12 A deadweight tester is used to test the accuracy of a pressure gauge. The piston is 25 mm in diameter, and when a mass of 9 kg is placed on it, the gauge reading is 210 kPa. What is the error in the gauge reading if the mass of the piston and carrier is 1 kg?

+5.1%

2.13 An external force is applied to a piston and cylinder (100 mm in diameter) and compresses the gas in it until the pressure reaches 1 MPa (abs) on a day when atmospheric pressure is normal.

(a) What is the maximum value of the external force?

(b) What pressure would a gauge attached to the cylinder show?

(c) If the compression took place at an elevation at which atmospheric pressure was one-half the sea-level value, what external force would now be required?

(d) If no atmosphere surrounded the piston, what force would be required?

(a) 7.06 kN   (b) 898.7 kPa   (c) 7.46 kN   (d) 7.85 kN

2.14 An aircraft's pressure gauge is calibrated at sea-level and is used to measure tank pressure. The reading is 500 kPa. If the tank pressure does not change, what reading will the gauge show when the aircraft is flying at an altitude at which atmospheric pressure has dropped to 80 kPa?

521.3 kPa

2.15 A water manometer is used to measure the pressure in a gas main. The graduated scale reads 238 mm in one limb and 112 mm in the other. Atmospheric pressure is normal. Determine:

(a) the gas pressure in kPa (gauge);

(b) the absolute gas pressure.

(a) 1.24 kPa   (b) 102.5 kPa

2.16 A manometer containing kerosene (*RD* 0.78) is attached to the gas main given in Problem 2.15. The limb exposed to the atmosphere reads 226 mm (above a datum). What is the reading on the limb connected to the gas main (above the same datum)?

    64.5 mm

2.17 A manometer is filled with liquid of *RD* 1.2. What will be the height difference in the limbs if it measures a pressure of:

(a) 106 kPa on a day when atmospheric pressure is normal;

(b) 99 kPa when atmospheric pressure is 763 mm of mercury? (*RD* mercury = 13.6.)

    (a) 399 mm   (b) 238 mm

2.18 A four-cylinder engine has a bore of 90 mm and a stroke of 100 mm. The clearance volume of each cylinder is 0.065 L. Assuming compression starts at the bottom of the stroke, determine:

(a) the capacity of the engine;

(b) the compression ratio of the engine.

    (a) 2.54 L   **(b)** 10.8:1

2.19 A four-cylinder engine has a bore of 100 mm, a stroke of 100 mm and operates on a compression ratio of 7.6:1. Determine the capacity of the engine and the clearance volume per cylinder, assuming that compression starts at the BDC position.

    It is desired to increase the compression ratio to 8.3:1 by planing the head of the engine. Determine the depth of cut necessary, assuming the head is cylindrical, with the diameter equal to the bore over this depth.

    3.14 L; 0.119 L; 1.45 mm

2.20 A double-acting diesel engine for a ship has eight cylinders with bore 1 m and stroke 1.3 m. The piston-rod diameter is 150 mm. Determine:

(a) the capacity of the engine;

(b) the distance between the top face of the piston and the top of the cylinder head in the TDC position and the distance between the bottom face of the piston and the bottom of the cylinder in the BDC position for a compression ratio of 18:1 on both sides of the piston. Assume compression starts at the extreme stroke positions and that the piston and cylinder heads are flat.

    (a) 16.15 m^3 (b) 76.5 mm; 76.5 mm (same)

CHAPTER 3

Energy

≡

Objectives

On completion of this chapter you should be able to:

- describe the various forms of energy of importance in thermodynamics;
- calculate the work done by a force or the work done from the pressure–volume diagram;
- calculate translational and rotational power;
- calculate sensible and latent heat transfers;
- calculate the energy available from the combustion of a fuel;
- describe the mechanism of nuclear energy (fission and fusion).

Introduction

In the past, much attention was given to the search for a perpetual-motion machine, that is, a machine that could do work and run forever. Numerous ingenious devices were proposed and constructed but each one failed. However, each failure was explained by some imperfection in the design or construction, which could be overcome by a yet more ingenious device. Only after repeated failures over many centuries came the realisation that work could not come from nothing and that some source of energy was required. Furthermore, it was realised that *the amount of work that could be done could not exceed the amount of energy converted.* Consequently, one of the most fundamental and important laws of science, the **principle of energy conservation** was formulated.

There are many forms of energy that are mutually convertible and can be measured in the same units. In this chapter, the forms of energy important in the study of thermodynamics are treated.

3.1 ENERGY

Energy may be defined as *the capacity to do work or to cause heat to flow.* In *system* terms, energy may be defined as *that property of a system that changes by an amount equal to the work or heat transferred across the system boundary.*

Important points about energy are listed as follows:

- *Energy can exist in many different forms.* Some forms of energy relevant to the study of thermodynamics are listed in Table 3.1.
- *In principle, all forms of energy are mutually convertible.*
- *There is a transfer or flow of energy when a change in form takes place.*
- *In thermodynamics, transfer of energy is usually heat or work.*
- *Heat and work cannot be stored as such,* and it is basically incorrect to use the terms 'stored heat' or 'stored work'.[*]
- *Stored energy is a property, whereas transferred energy is not.* This means that whenever heat flows or work is done, the amount transferred depends on the method or process of transfer as well as the end states. On the other hand, stored energy has a value that is independent of the process.
- *Energy is a scalar quantity with no direction relative to a frame of reference.* However, when energy is transferred across a system boundary, it is necessary to know whether the energy transfer is *into* or *out of* the system. This may be specified by use of a sign convention.
- *There is no creation or destruction of energy when a change in form takes place,* so that the total energy remains constant before, during and after the change. This is known as the **principle (or law) of conservation of energy** or the **first law of thermodynamics**.
- *All forms of energy have the same units* in the SI system, which are the units of work, namely Nm or J. A common (but non-SI) unit of energy used by electrical-power-generating authorities is the kilowatt hour (kWh). This is the amount of electrical energy used by a one kilowatt appliance in one hour. Because 1 h = 3600 s, 1 kWh = 1 kW × 3600 s = 3600 kJ = 3.6 MJ.
- *Although there is no change in quantity when a change in form takes place, it is usually not possible to obtain the form required without, at the same time, converting some energy into forms not required or unusable.* Energy not required or unusable is often called lost energy. More correctly, it is wasted energy or unusable energy.
- The efficiency of an energy-conversion process has already been defined in Chapter 1, namely:

$$
\boxed{
\begin{aligned}
\eta &= \frac{\text{usable energy output}}{\text{energy input}} \\
&= \frac{\text{energy input} - \text{energy loss}}{\text{energy input}}
\end{aligned}
}
\qquad \text{efficiency of energy conversion} \quad (3.1)
$$

which may be expressed as a fraction or a percentage, and is dimensionless, that is, it has no units.

---

[*] This point often causes confusion. For example, when a mass of water is heated, it *appears* as if heat is being stored, and in fact one form of hot-water heater is known as a 'storage heater'. What is really happening is that when a mass of water is heated, the heat is converted into internal energy of the water molecules and stored as such. This is analogous to the way in which electrical energy can be converted to chemical energy and stored as such in a battery.

Table 3.1   *Some forms of energy*

Classification	Form	Type
mechanical energy	• potential energy	stored
	• kinetic energy	stored
	• work	flow
heat energy	• sensible heat	flow
	• latent heat	flow
chemical energy	• molecular bonds	stored
internal energy	• molecular motion	stored
electrical energy	• current flow	flow
nuclear energy	• atomic bonds	stored

3.2 *POTENTIAL ENERGY*

Gravitational potential energy (usually called simply 'potential energy', *PE*) *is the energy possessed by a mass due to the force of attraction of the earth when the mass is located at some height h above any arbitrary datum*, and is given by:

$$PE = mgh$$

potential energy (3.2)

where m = mass (kg)
g = gravitational constant (N/kg or m/s^2)
h = vertical height (m)
PE = potential energy (J)

Potential energy is often of relatively small magnitude compared with the other energy forms, and can often be neglected. However, it is a form of energy that stores very well and does not degrade with time.

3.3 *KINETIC ENERGY*

Kinetic energy (KE) is energy due to motion. It may be linear if the motion is linear, or rotational if the motion is circular.[*]

The equation for linear kinetic energy is:

$$KE = \tfrac{1}{2}mv^2$$

translational kinetic energy (3.3)

---

[*] It is possible (and indeed common) for both modes to occur simultaneously. For example, a rolling ball has linear kinetic energy due to the linear motion of the centre of mass, and rotational kinetic energy due to the rotation about the centre of mass.

where m = mass (kg)

   v = velocity (m/s)

   KE = kinetic energy (J)

Like potential energy, kinetic energy may not be of great significance in thermodynamics unless there is a large mass-flow rate or velocity change in a system (such as in jet engines and rockets) and it can often be neglected. Unlike potential energy, however, kinetic energy does not store very well since the frictional losses that usually occur with motion degrade it.

3.4   *WORK*

Like kinetic energy, work (W) may be *linear* or *rotational*. *Linear work is the product of a constant force and the distance moved by that force along its line of action.* Mathematically:

$$\boxed{W = Fx}$$ linear work (3.4)

where F = the constant force (N) in any given direction

   x = distance (m) moved in that direction

   W = work (Nm or J) done by the force F

Variable force

In many applications, the force varies while moving through distance x, so the equation should be written in differential form as:

$$dW = Fdx$$

where force F may be considered as being constant only over an infinitesimally small distance, dx. The total work done is then:

$$W = \int_{x_1}^{x_2} Fdx$$

Rotational work

In many cases in engineering, work is done by a force moving about a centre of rotation as occurs with a rotating shaft or flywheel. Here the motion is rotational (circular) rather than linear. The equation for rotational work is equivalent to the equation for linear work except that torque is used instead of force and angular displacement is used rather than linear distance. That is:

$$\boxed{W = T\theta}$$ rotational work (3.5)

where T = torque or turning moment, which is defined as the *tangential force multiplied by the radius at which the force acts*. Because force has units newtons, and radius has units metres, the units of torque are Nm.

θ = angular displacement in radians. Note that a displacement of one revolution (360°) = 2π radians.

 ## *Self-test problem 3.1*

A mass of 5 t is lifted 20 m vertically off the ground to a rest position.
(a) Determine the work done.
(b) Determine the change in potential energy.

The mass is then dropped.
(c) Determine the velocity of the mass after it has descended 5 m assuming that all the potential energy is converted to kinetic energy.

Relationship between work, pressure and volume

In many applications such as in heat engines, compressors and pumps, the applied or opposing force is due to fluid pressure. In such cases, the work transfer may be obtained as follows:

Consider a piston and cylinder as shown in Figure 3.1. The force due to fluid pressure on the face of the piston is just sufficient to overcome the opposing forces due to the applied force F, the friction and the atmospheric pressure, so that the piston moves a distance x from position ① to position ②. During this motion the pressure p (absolute) acting on the face of the piston remains constant. In practice, this can occur because more fluid flows into the cylinder as shown in Figure 3.1 or, if the fluid is enclosed, it can occur by heating and expanding the fluid.

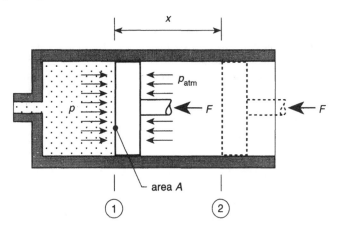

Fig. 3.1   *Constant-pressure displacement of a fluid*

Force on the face of the piston due to fluid pressure is pA. The work done by the fluid is pAx. But $Ax = V$, the displaced volume $(V_2 - V_1)$, so

$$\boxed{W = pV = p(V_2 - V_1)}$$ work for a constant-pressure process (3.6)

where p = the constant fluid pressure (absolute) (Pa)
           V = displaced volume (m^3)
           W = work done (Nm or J) *by the fluid*
           V_1 = initial volume (m^3)
           V_2 = final volume (m^3)

If the pressure does not remain constant, Equation 3.6 is written in differential form:

$$dW = p\,dV$$
$$W = \int_{V_1}^{V_2} p\,dV \quad \text{work for a variable-pressure process (3.7)}$$

Before the integration can be performed to find the total work done, it is necessary to obtain a relationship between p and V. Note the following important points:

- Equations 3.6 and 3.7 are for the work done *by the fluid*, that is, the work done on the face of the piston. They are not equations for the available work or output work because atmospheric pressure and friction act to oppose motion. This is clarified in Example 3.2.
- V_1 is the initial volume and V_2 is the final volume. (This is easy to remember if you think of the 1 as standing for 'i'.)
- For an *expansion*, work is done *by* the fluid:
  $V_2 > V_1$, so $V_2 - V_1$ is *positive* and W is *positive*.
  For a *compression*, work is done *on* the fluid:
  $V_2 < V_1$, so $V_2 - V_1$ is *negative* and W is *negative*.
  This is the sign convention most used in thermodynamics and stems from the fact that heat engines produce work by the expansion of a fluid, so *work out of a system* is *positive*.

Example 3.1

A hot-air balloon contains a volume of 20 m^3 of air at atmospheric pressure. The burners are now turned on and the balloon inflated to 1020 m^3. If the pressure inside the balloon can be considered to remain essentially constant, how much work is done during the expansion?

Solution

Since this is a constant-pressure process:

$$W = p(V_2 - V_1)$$
$$= 101.3 \times 10^3 \times (1020 - 20)$$
$$= 101.3 \times 10^6 \text{ (Nm or J)}$$
$$= \textbf{101.3 MJ}$$

3.5 *PRESSURE–VOLUME DIAGRAM*

Suppose for the constant-pressure expansion process illustrated in Figure 3.2, a graph were drawn of gas pressure versus volume. This graph is often used in thermodynamics and is known as a *p–V* **diagram**. The diagram would look as shown in Figure 3.2. The crosshatched area is:

$$p(V_2 - V_1) = pV,$$

which is the same expression as was obtained previously for the work done.

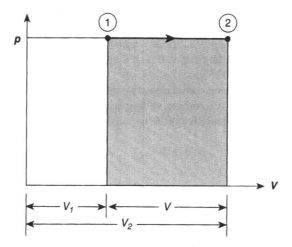

Fig. 3.2 *p–V diagram for constant-pressure expansion*

If the pressure changes as the piston moves (volume expands), the diagram might look as shown in Figure 3.3. Now the crosshatched area under the curve is:

$$\text{area} = \int_{V_1}^{V_2} p \, dV$$
$$= W \text{ (work done by the fluid)}$$

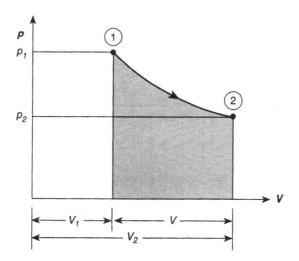

Fig. 3.3 *p–V diagram for a general expansion process*

Hence the important conclusion follows: *the area under the p–V curve is the work done by the fluid.* This is a very useful relationship, which explains why the *p–V* diagram is so widely used in thermodynamics.

Note that in order to determine the work done, it is not necessary to have a mathematical relationship between pressure and volume. If the diagram can be drawn (say from experiment), the area may be measured and the work determined.

Example 3.2

A gas in a cylinder of bore 50 mm is at a pressure of 140 kPa (absolute). The gas is heated and expands at constant pressure moving the piston a distance of 300 mm. Sliding friction force is 20 N. Draw a neat p–V diagram and determine:

(a) the work done by the gas;
(b) the work done in overcoming air resistance;
(c) the work done in overcoming sliding friction;
(d) the nett work available from the expansion.

Solution

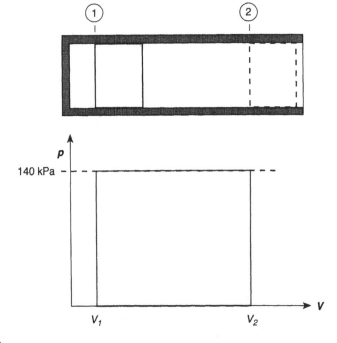

Fig. 3.4

(a) Since this is a constant-pressure process,

$$W = p(V_2 - V_1) = pAx$$

$$\text{Now } A = \pi \times \left(\frac{0.05}{2}\right)^2 = 1.963 \times 10^{-3} \text{ m}^2$$

$$\therefore W = 140 \times 10^3 \times 1.963 \times 10^{-3} \times 0.3$$
$$= \textbf{82.5 J}$$

(b) Similarly, the work done in overcoming air resistance is:

$$W = 101.3 \times 10^3 \times 1.963 \times 10^{-3} \times 0.3$$
$$= \textbf{59.7 J}$$

(c) The work done in overcoming friction is:

$$W = Fx$$
$$= 20 \times 0.3$$
$$= \textbf{6 J}$$

(d) Hence, the nett work available is:

$$82.5 - 59.7 - 6 = \textbf{16.8 J}$$

Work for a cycle

When a series of processes is performed in such a way that the system returns to its original state, then the p–V diagram will form a closed loop. In this case, the nett work per cycle done by the fluid is represented by the area enclosed by the loop on the p–V diagram as shown crosshatched in Figure 3.5.

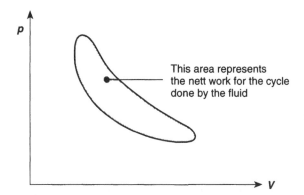

Fig. 3.5 *For a cycle, the nett work done by the fluid is represented by the enclosed area*

 Self-test problem 3.2

A gas expands from a volume of 0.2 m³ and a pressure of 300 kPa to a volume of 0.7 m³ and a pressure of 100 kPa, such that the p–V diagram is a straight line. Draw a p–V diagram (to scale) and determine the work done by the gas during the expansion.

3.6 *POWER*

The amount of work performed by a machine does not indicate the size or capacity of the machine unless there is a time frame. Any machine is capable of doing any amount of work provided it runs long enough. The more quickly the work is to be performed, the more powerful the machine required and the greater the rate of energy conversion. *Power (P) is the time rate of transfer of work or of conversion of energy:*

$$P = \frac{W}{t} = Fv$$

power (linear) (3.8)

where W = work (Nm or J)
t = time taken (s)
F = force (N)
v = velocity (m/s)
P = power (J/s or W)

The watt is a small unit of power, so in engineering the kilowatt (kW) and megawatt (MW) are frequently used.

Shaft power

Often, mechanical power is input or output by means of a rotating shaft. In such cases torque (turning moment) is exerted on the shaft rather than force and the velocity is angular rather than linear. The rotational equation for shaft power is:

$$\boxed{P = T\omega}$$ power (rotational) (3.9)

where T = torque (Nm) = Fr (F = force (N), r = radius (m))
ω = rotational velocity (rad/s)

If the rotational velocity is in revolutions per minute, N (rpm), the conversion is:

$$\boxed{\omega = \frac{2\pi N}{60} = \frac{\pi N}{30}}$$ angular velocity (3.10)

Example 3.3

A p–V diagram taken from a twin-cylinder reciprocating gas compressor shows a nett work per cylinder of 2.5 kJ. The compressor runs at 420 rpm and each cylinder completes one cycle per revolution of the crankshaft.

Determine the average torque and power required to drive the compressor if frictional losses are 14% of the input power.

Solution

Because there are two cylinders, the work done on the gas (fluid work) for one revolution of the crankshaft is:

$$W = 2 \times 2.5 = 5 \text{ kJ}$$

Time taken for one revolution of the crankshaft is $t = \dfrac{60}{420} = 0.143$ s

$$P = \frac{W}{t}$$
$$= \frac{5}{0.143}$$
$$= 35 \text{ kW}$$

This is the fluid power.

Since $\eta = \dfrac{\text{energy out}}{\text{energy in}}$ and losses are 14%, $\eta = 86\%$.

$$\therefore P_{in} = \frac{\text{fluid power}}{\eta} = \frac{35}{0.86}$$

$$= \textbf{40.7 kW}$$

$$\omega = \frac{\pi N}{30} = \pi \times \frac{420}{30} = 44 \text{ rad/s}$$

Since $P = T\omega$, then

$$T = \frac{P}{\omega} = \frac{40\,700}{44}$$

$$= \textbf{925 Nm}$$

Self-test problem 3.3

An air motor develops a torque of 3.5 Nm at the output shaft when the speed is 4500 rpm. Mechanical losses in the motor are 10%. Determine:

(a) output power;
(b) input fluid power.

3.7 *HEAT*

Heat (Q) is a form of energy that occurs only as a flow (or transfer) of energy as a result of a temperature difference. *The direction of heat flow in a substance or across a single boundary of a system is always from the higher to the lower temperature.* This statement is known as the **second law of thermodynamics**. As heat is a form of energy, it also has units, joules, although kilojoules or megajoules are used more often.

Note that when heat flows into or out of a substance, the temperature of that substance does not necessarily change. If the heat flow *does* cause change in temperature, this is known as **sensible heat**. If there is *no* change in temperature but rather a change in phase, this is known as **latent heat**. Examples of sensible heating are the rise in temperature of water in a hot-water heater, heating a block of metal in a gas flame or the heating of the road surface on a hot day. However, if water boils or metal melts, these changes in state are examples of latent heat.

If the temperature change is not too large, it has been found that for any given material, the sensible heat flow is given by:

$$\boxed{Q = mc\Delta T = mc(T_2 - T_1)} \qquad \text{sensible heat (3.11)}$$

where Q = quantity of heat transferred (J)
    m = mass of the material (kg)
    c = specific heat capacity (J/kgK)
    T_1 = initial temperature (°C or K)
    T_2 = final temperature (°C or K)
    ΔT = temperature difference (°C or K) = $T_2 - T_1$

Notes

1   It does not matter if T_2 and T_1 are in °C or K, provided they are both expressed in the same unit, since the difference $T_2 - T_1$ will be the same in both cases.

2   T_1 is the *initial* temperature and T_2 is the *final* temperature. Therefore if the system *gains* heat, $T_2 > T_1$ so $T_2 - T_1$ is positive, and Q is *positive*. Conversely, if a system *loses* heat, $T_2 < T_1$ so $T_2 - T_1$ is negative and Q is *negative*.

3   The meaning of specific heat capacity follows from Equation 3.11, that is, *specific heat capacity is the amount of heat (in joules) required to raise the temperature of 1 kg of a substance by 1°C (or K)*.

4   Equation 3.11 assumes constant specific heat capacity while heat is being transferred. In fact, experiments show that for most substances specific heat capacity changes with temperature. However, Equation 3.11 may still be used with an *average* value of specific heat capacity over the temperature range involved.

   In this book, average values of specific heat capacity given in Appendix 4 are used in all problems unless stated otherwise.

Example 3.4

(a) How much heat is required to raise the temperature of a 500 g copper cup from 20°C to 80°C?

(b) If the copper cup had contained 300 g of water, how much heat would have been required for the same temperature rise?

(c) If in part (b) the heat energy is provided by an electrical immersion heater (100% efficient), what is the energy cost if electricity costs 11 cents per kWh?

Solution

(a) From Appendix 4, c for copper = 0.39 kJ/kgK

$$Q = mc(T_2 - T_1)$$
$$= 0.5 \times 0.39 \times (80 - 20) \text{ kJ}$$
$$= \textbf{11.7 kJ}$$

(b) The heat required is the sum of the heats required for each component separately. For the cup:

$$Q = 11.7 \text{ kJ}$$

For the water:
from Appendix 4, c for water = 4.19 kJ/kgK

$$Q = mc(T_2 - T_1)$$
$$= 0.3 \times 4.19 \times (80 - 20) \text{ kJ}$$
$$= 75.4 \text{ kJ}$$

Total quantity of heat required is:

$$11.7 + 75.4 = \textbf{87.1 kJ}$$

(c)   1 kWh = 3.6 MJ = 3600 kJ

$$\therefore 87.1 \text{ kJ} = \frac{87.1}{3600} \text{ kWh} = 0.0242 \text{ kWh}$$
$$\therefore \text{cost} = 11 \times 0.0242$$
$$= \textbf{0.266 cents}$$

Latent heat

No temperature change is involved during a phase change so the equation for latent heat does not include temperature terms and is:

$$Q = mL$$ latent heat (3.12)

where L is the latent-heat capacity, usually known simply as 'latent heat' (J/kg). There are two phase changes possible, namely the solid–liquid change known as **fusion**, and the liquid–vapour change known as **evaporation**. The latent heat is different in each case; for example, for water at atmospheric pressure:

      latent heat of evaporation = 2257 kJ/kg
      latent heat of fusion   = 335 kJ/kg

Example 3.5

How much additional heat is required to boil away one-half of the water from the cup in Example 3.4?

Solution

Additional *sensible* heat for copper (to bring to 100°C) = $0.5 \times 0.39 \times (100 - 80)$
= 3.9 kJ
Additional *sensible* heat for water (to bring to 100°C) = $0.3 \times 4.19 \times (100 - 80)$
= 25.1 kJ
Latent heat required for water = $0.5 \times 0.3 \times 2257$
= 338.6 kJ
Total additional heat required = 3.9 + 25.1 + 338.6
= **367.6 kJ**

 ## *Self-test problem 3.4*

A 600 g block of ice is at –20°C. 250 kJ of heat are now transferred to the ice, which all melts. Determine the final temperature of the water. Use latent heat values for water previously given and use Appendix 4 for specific heat capacities.

3.8 CHEMICAL ENERGY

A chemical reaction occurs when the molecules of a substance dissociate or combine with other molecules to form new substances. The regrouping of the atoms involves energy known as chemical energy.

*Energy is released if the reaction is **exothermic**, or absorbed if the reaction is **endothermic**.* For example, under normal conditions, water is stable. However, if energy is added, say by electrolysis, the water molecules dissociate into hydrogen and oxygen. Hydrogen and oxygen are themselves stable but may readily be recombined by combustion into water again. During the combustion process, the same energy is released as was originally required to separate the molecules. The chemical equations for the process are

$$2H_2O + energy \rightarrow 2H_2 + O_2$$
$$2H_2 + O_2 \rightarrow 2H_2O + energy$$

The amount of heat energy liberated by combustion of a substance is known as the **energy content** or **heating value** of the substance. (It was originally known as 'calorific value' because heat was measured in calories.)

The energy available by combustion of a substance is given by:

$$\boxed{Q = mE}$$ combustion heat energy (3.13)

where E = energy content of the substance (J/kg)
         m = mass of substance burnt (kg)
         Q = heat energy liberated (J)

Note that if the energy content is given in kJ/kg or MJ/kg the heat energy will accordingly be in kJ or MJ.

Example 3.6

How much coal must be burnt in order to raise the temperature of 1200 L of water from 15°C to 90°C if 50% of the heat of combustion is utilised? The energy content of the coal is 31 MJ/kg.

Solution

$$T_1 = 15°C$$
$$T_2 = 90°C$$
$$c = 4.19 \text{ kJ/kgK}$$
$$m = 1200 \text{ kg}$$
$$Q = mc(T_2 - T_1)$$
$$= 1200 \times 4.19 \times (90 - 15) \text{ kJ}$$
$$= 377.1 \text{ MJ}$$

Since only 50% of the heat is utilised, the total amount of heat required from the fuel is

$$377.1 \times 2 = 754.2 \text{ MJ}$$

$$\text{Amount of coal required } m = \frac{754.2}{31}$$
$$= \textbf{24.33 kg}$$

 ## Self-test problem 3.5

How many litres of fuel need to be supplied per hour to an internal-combustion engine running on petrol and producing 80 kW of power if the efficiency of the engine is 33%. The energy content of petrol is 46.5 MJ/kg and the relative density is 0.74.

3.9 INTERNAL ENERGY

Internal energy (U) is the energy possessed by the molecules of a substance. Internal energy was discussed in Chapter 2 in connection with molecular movement. It is difficult to determine internal energy on an absolute scale but changes in internal energy can be determined by the energy transfers that cause changes in the internal energy of a substance. This is discussed further in Chapter 4.

3.10 *NUCLEAR ENERGY*

Nuclear energy (sometimes called atomic energy) results from a change in the nucleus of the atom caused by a nuclear reaction. Nuclear reactions may be exothermic or endothermic. Exothermic reactions, which liberate energy, may be of two types:

- **fission**, where the nucleus breaks down into a number of smaller nuclei;
- **fusion**, where small nuclei combine to form a larger nucleus.

In all nuclear reactions, the combined law of conservation of matter/energy applies, so that

$$[\text{matter} + \text{energy}] \text{ before} = [\text{matter} + \text{energy}] \text{ after}$$

where the equivalence between matter and energy is expressed by the Einstein equation:

$$\boxed{E = mc^2}$$ nuclear energy (3.14)

where m = mass (kg)
E = energy (J)
c = speed of light (m/s)

Because the speed of light is so large (about 3×10^8 m/s) a huge amount of energy is released by the conversion of a small amount of mass into energy. For example, the fusion reaction of hydrogen in the sun liberates *about four million times* the energy that would be obtained from combustion of the same amount of hydrogen!

 ## *Self-test problem 3.6*

A large power station produces 800 MW of power continuously with an efficiency of 40%. Determine:

(a) the amount of coal, of energy content 27 MJ/kg, needed per day if the power station is a coal-burning one;
(b) the amount of mass that would need to be converted into energy per day if the power station were a nuclear one.

 ## Summary

Energy is the capacity to do work or to cause heat to flow. Energy exists in many different forms, all with the same units (J) but some that may be stored and some that exist only as a *flow* or *transfer* of energy. Two most important energy transfers in engineering occur as work transfer and heat flow.

Work transfer is often a result of pressure and volume changes to a fluid. In such cases, the work transfer is equal to the area under the p–V diagram of the process. An important special case is the constant-pressure process, which is a horizontal straight line on the p–V diagram. The work done for this process is given by $W = p(V_2 - V_1)$. By convention, work out of a system is considered to be positive. The time rate of transfer of work is known as power and may be translational ($P = Fv$) or rotational ($P = T\omega$). The unit of power is the Watt (J/s).

> Heat flow occurs as a result of a temperature difference and may be sensible (causes a change in temperature) or latent (causes a change in phase or state). Sensible heat flow is given by $Q = mc(T_2 - T_1)$ and latent heat flow by $Q = mL$. By convention, heat into a system is considered to be positive.
>
> Much of the energy used by humans comes from combustion, which is a conversion of chemical energy into heat energy. The amount of energy available is known as the *energy content E* (formerly calorific value). For a mass of fuel, m, the heat energy released is given by $Q = mE$.
>
> Nuclear energy is derived from changes to the structure of the nucleus of atoms and may be a combination of nuclei (fusion) or a breakdown of nuclei (fission). Whereas the energy of the sun is a result of fusion, nuclear reactions on earth are all of the fission type. Because of the differing masses of nuclei of different types, a nuclear reaction changes the mass of the atoms and the amount of energy released is given by the Einstein equation $E = mc^2$.

Problems

Note Where required in these problems, use specific heat capacities given in Appendix 4. For water, latent heat of evaporation = 2257 kJ/kg and latent heat of fusion = 335 kJ/kg.

3.1 (a) Explain the distinction between *temperature, internal energy,* and *heat. Refer to the molecular model in your answer.*

(b) Explain the distinction between *latent* heat and *sensible* heat.

(c) State the *second law of thermodynamics.*

(d) Explain what is meant by *chemical energy.*

(e) How does a chemical reaction differ from a nuclear reaction?

3.2 A car of mass 1.2 t descends a gradient of 1 in 20 (sine). Initially the road speed is 30 km/h. After moving down the gradient for 1 km the road speed is 80 km/h. Determine:

(a) the change in potential energy;

(b) the change in kinetic energy;

(c) the amount of heat dissipated to the atmosphere as a result of the descent.

　　(a) –588.6 kJ (loss)　(b) 254.6 kJ (gain)　(c) 334 kJ

3.3 A mass of 5 kg is moved a distance of 20 m at a velocity of 4 m/s along a horizontal surface, the coefficient of friction being 0.4 between the surfaces in contact. Determine:

(a) the heat generated at the surfaces;

(b) the power required to overcome friction.

　　(a) 392.4 J　(b) 78.5 W

3.4 An ice box contains 2 kg of ice at 0°C. If heat is leaking into the box at a rate of 12 kJ/h, calculate how long the ice will last.

　　55.8 h

3.5 A shaft of diameter 150 mm running at 500 rpm is supported by journal bearings, which have a coefficient of friction of 0.02. The load on each bearing is 20 kN. Calculate, for each bearing:

(a) the power dissipated by friction;

(b) the heat generated per minute.

　　(a) 1.57 kW　(b) 94.2 kJ

3.6 Determine the oil flow per minute required for each journal bearing given in Problem 3.5 if the temperature rise of the oil is not to exceed 60°C. For the oil, $c = 1.8$ kJ/kgK, $RD = 0.85$.

  1.03 L/min

3.7 A pump discharges 300 L of water per minute through an elevation of 20 m. The overall efficiency of the pumping system is 70%. Determine:
(a) the input power required to raise the water;
(b) the annual cost of running the pump if it operates an average of 12 h per day, 7 days per week, and electricity costs 8c per kWh.

  (a) 1.4 kW **(b)** $490

3.8 A *p*–*V* diagram taken from a two-stroke single-cylinder air motor shows a nett work of 25 J when the motor runs at 5600 rpm. Determine the average shaft power and torque if losses are 25%. *Note* A two-stroke motor gives one power stroke per revolution.

  1.75 kW; 2.98 Nm

3.9 In Hero's aeolipile illustrated in Figure P3.9, the reaction force developed at each nozzle is 4 N when it spins at 400 rpm. Determine the resistive torque and power (due to friction and air resistance).

  1.76 Nm; 73.7 W

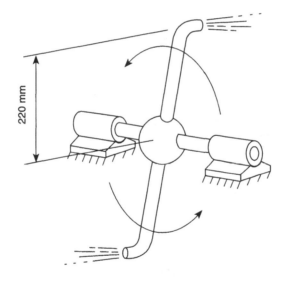

Fig. P3.9

3.10 An electric furnace has dimensions 2 m × 2 m × 3 m. When it reaches forging temperature of 1200°C, a 2 t block of steel at 20°C is placed in the furnace. If there is a heat loss of 300 W/m² from all surfaces, and a constant energy input of 20 kW, calculate how long it will take before the steel can be forged.

  29 h

3.11 In an industrial process, a mixer transfers energy at a rate of 5 kW into 500 L of liquid, which has a specific heat capacity of 3.5 kJ/kgK and a relative density of 0.92. Determine the increase in temperature of the liquid per minute during mixing if:
(a) there is no heat loss from the mixing tank;
(b) the heat loss from the tank is 40% of the input energy.

  (a) 0.186°C **(b)** 0.112°C

3.12 How high would a mass have to be lifted in order to expend the same amount of energy as would be required to heat an equal mass of water through 1°C? (Think about the answer; it gives insight as to why a small amount of heat energy can perform a great deal of mechanical work.)

   427 m

3.13 An aeroplane of mass 5 t takes off, and after 30 min has climbed to an altitude of 3000 m and has reached a cruising speed of 700 km/h. Calculate the potential and kinetic energy of the aircraft (neglecting aircraft-mass change due to consumption of fuel). If the aircraft consumes, per second, 0.1 kg of fuel of energy content 46 MJ/kg during the ascent, calculate what percentage of the chemical energy of the fuel has been converted to potential energy and what percentage to kinetic energy.

   To what form of energy has the remainder of the chemical energy of the fuel been converted?

   147 MJ; 94.5 MJ; 1.78%; 1.14%

3.14 A cast-iron saucepan of mass 1.4 kg is filled with 5 L of water at 15°C, placed on an electric hotplate, and brought up to boiling point. What is the cost of the energy used if electricity costs 10 cents per kWh and the hotplate has an efficiency of 70%?

   7.3 cents

3.15 In an ice-making plant, ice at −5°C is produced from water at 20°C. What is the refrigeration capacity (heat removal capacity) of the plant per hour if the plant can produce 6000 kg of ice in an 8 h shift?

   321.75 MJ/h

3.16 An aluminium beer can has a mass of 14 g and contains 375 mL of beer. How much ice at 0°C will have to melt in order to cool 24 full cans from 25°C to 2°C? The specific heat of beer is 4.0 kJ/kgK and relative density is 0.985. Also calculate the error introduced by neglecting the mass of the can itself.

   2.4 kg; 0.83%

3.17 A car of mass 1 t travels at a constant speed of 60 km/h. The resistance force at this speed is 10% of the vehicle weight.

   (a) Calculate the power being delivered at the vehicle's wheels.
   (b) If the transmission system has an efficiency of 80%, calculate the engine power output.
   (c) If the engine itself has an efficiency of 28%, calculate the fuel consumption in litres of petrol per 100 km. The energy content of petrol is 45 MJ/kg and the *RD* is 0.7.

   (a) 16.35 kW   (b) 20.44 kW   (c) 13.9 L/100 km

3.18 A gas of volume 4 m^3 at pressure 1 MPa (abs.) is heated at constant pressure to a volume of 7 m^3. Draw a p–V diagram for this process and determine the work done.

   3 MJ

3.19 A gas is compressed in a cylinder so that the pressure rises linearly from 100 to 600 kPa (abs.) while the volume falls from 0.06 m^3 to 0.02 m^3. The gas is then cooled to a final volume of 0.01 m^3 at a constant pressure of 600 kPa (abs.). Draw a p–V diagram and calculate the work done.

   −20 kJ

3.20 A reciprocating engine has a bore of 60 mm and a stroke of 60 mm, and runs at 3500 rpm. During the working stroke, the average gas pressure on the face of the piston is 800 kPa (gauge) and the sliding friction is 30 N. Determine the following for the working stroke:

(a) the work done by the gas;

(b) the work done in overcoming air resistance;

(c) the work done in overcoming sliding friction;

(d) the nett work available.

     (a) 152.9 J   (b) 17.2 J   **(c)** 1.8 J   **(d)** 133.9 J

3.21 A two-stroke single-cylinder reciprocating engine runs at 3500 rpm. A *p–V* diagram for the gas in the cylinder shows a nett work per cycle of 88 J. Friction and air-resistance losses reduce the amount of power available at the crankshaft by 15%. Determine:

(a) the nett power of the gas (fluid power);

(b) the power at the crankshaft (output power);

(c) the output torque.

     (a) 5.13 kW   **(b)** 4.36 kW   (c) 11.9 Nm

3.22 The energy content of black coal (by combustion) is about 30 MJ/kg. How much black coal would need to be burnt to release the same amount of energy as the conversion of a 1 kg mass of this coal into energy? The speed of light is 3×10^8 m/s.

     3×10^6 t

3.23 A nuclear power station has a power output of 1000 MW. Nuclear fission of 1 kg of U^{235} releases 90×10^6 MJ. U^{235} is about 0.71% of natural uranium U^{238}. The power station operates at full capacity for a year and has an efficiency of 40%.

(a) Determine how much natural uranium is required per annum.

(b) If the combustion energy content of black coal is 30 MJ/kg, how much coal would be required to produce the same power output by combustion (assuming the same efficiency)?

     (a) 123.4 t   (b) 2.63×10^6 t

Closed and open systems ≡

Objectives

On completion of this chapter you should be able to:

- define a closed and an open system and explain the meaning of steady flow in an open system;

- solve problems involving closed systems such as a fluid being heated or cooled while being stirred using the non-flow energy equation, and solve calorimetry problems, that is, problems involving internal heat exchanges only;

- describe the operation and use of a bomb calorimeter and determine the energy content of a solid or liquid fuel from typical test results;

- determine fluid velocity or mass or flow-volume flow rates in an open system such as a jet engine using the continuity equation;

- solve problems involving open systems such as turbines, boilers (steam generators), or heat exchangers using the steady-flow energy equation (without changes in potential or kinetic energy);

- describe the operation and use of a gas calorimeter and determine the energy content of a gaseous fuel from typical test results.

Introduction

In this chapter, closed and open systems are defined, and the equations applicable to them developed. Their importance lies in the fact that many devices encountered in thermodynamics, such as compressors, turbines, motors, engines, boilers, heat exchangers, refrigerators or air-conditioners, may be treated by closed-system or open-system analysis.

There are two important factors to consider, namely mass transfer and energy transfer. In this chapter, these are treated using black-box analysis, except in closed-system cases when it is sometimes necessary to look inside the system to analyse the energy transfers.

The energy transfers considered are heat and work. Other possible energy transfers, such as electrical energy, are not considered, but the equations developed are valid in such cases provided the energy is treated as an equivalent quantity of heat. It is important remember the sign convention for heat and work already stated in Chapter 3, namely that *heat in* is *positive*, and *work out* is *positive*.

4.1 CLOSED SYSTEM

A closed system is one in which there is no mass transfer across the system boundary. The mass within the system therefore remains constant, so the closed system is also known as the **control-mass system**. Two examples of a closed system are illustrated in Figure 4.1.

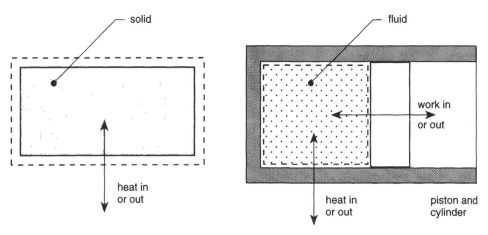

Fig. 4.1 *Closed systems*

Note Solids are rigid and therefore do not need to be confined in a vessel or container to be treated as a closed system. However, gases need to be confined in a vessel or container to prevent mass transfer. A liquid in an open vessel may often be treated as a closed system unless evaporation occurs.

4.2 NON-FLOW ENERGY EQUATION AND ITS APPLICATIONS

In the general case of a *closed* system, both heat and work can transfer simultaneously across the system boundary as shown in Figure 4.2 on page 68.

Usually with closed systems there is an insignificant change of potential or kinetic energy of the substance in the system. Hence the only change in energy of the system is that of internal energy. If the initial internal energy of the system was U_1 and the final internal energy was U_2, conservation of energy gives:

$$U_1 + Q - W = U_2$$

Note the negative sign for work because work–energy transfer has the opposite sign to heat–energy transfer. Therefore:

$$\boxed{Q - W = U_2 - U_1}$$ non-flow energy equation (4.1)

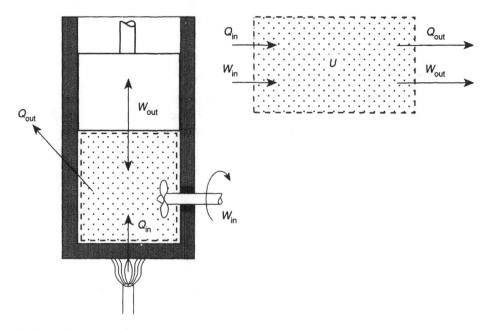

Fig. 4.2   *General energy transfers in a closed system*

If no phase change occurs, then over a restricted temperature range in which the specific heat capacity remains approximately constant:

$$\boxed{U_2 - U_1 = mc(T_2 - T_1)}\ \text{change in internal energy (4.2)}$$

Note   In the case of gases, c is the specific heat capacity at constant volume (written c_v). This is discussed further in Chapter 5.

Example 4.1

In a closed system, 50 kJ of heat flow in during a process in which 40 kJ of work are done. If heat losses are 20 kJ, calculate the change in internal energy during the process.

Solution

Nett heat in:

$$Q = 50 - 20$$
$$= + 30 \text{ kJ}$$

Work out:

$$W = +40 \text{ kJ}$$
$$Q - W = U_2 - U_1$$
$$30 - 40 = U_2 - U_1$$
$$U_2 - U_1 = \mathbf{-10 \text{ kJ}}$$

The minus sign signifies that there has been a drop in internal energy, that is, $U_2 < U_1$.

Example 4.2

In an industrial process, as shown in Figure 4.3, 800 L of a liquid, with specific heat capacity 2.6 kJ/kgK and a relative density of 0.88, are heated in a closed tank at the same time as it is being stirred. The heater output is 6 kW and the stirrer shaft power is 4 kW. Heat losses from the tank are 1.5 kW. The process continues for half-an-hour. The heat capacity of the tank itself is negligible. Determine:

(a) the change in internal energy of the liquid;

(b) the rise in temperature of the liquid if no phase change occurs.

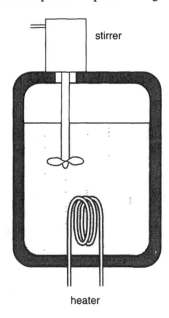

stirrer

heater

Fig. 4.3

Solution

(a) A 30-minute period is $30 \times 60 = 1800$ s. The energy transfers are as follows:

Nett heat flow rate $= 6 - 1.5 = 4.5$ kW (+ because heat in)
$Q = 4.5 \times 1800$ (kJ) $= 8.1$ MJ
Work transfer $= -4 \times 1800$ (kJ) $= -7.2$ MJ (– because work in)

Applying the non-flow energy equation:

$$Q - W = U_2 - U_1$$
$$\therefore 8.1 - (-7.2) = U_2 - U_1$$
$$\therefore U_2 - U_1 = \textbf{15.3 MJ}$$

(b) Since the substance is a liquid and there is no phase change, the internal energy change is given (very closely) by Equation 4.2:

$$U_2 - U_1 = mc\Delta T$$

The mass of liquid is $m = 800 \times 0.88 = 704$ kg

$$\therefore 15.3 \times 10^3 = 704 \times 2.6 \times \Delta T \text{ kJ}$$
$$\therefore \Delta T = \textbf{8.36°C}$$

Self-test problem 4.1

A stirrer of shaft power 500 W is used to stir 100 kg of water that is being heated. The heater output power is 3.5 kW and heat losses are 800 W. The initial temperature of the water is 20°C. Determine:

(a) how long it will take before the water reaches a temperature of 80°C;

(b) the internal energy change of the water.

Neglect the specific heat capacity of the vessel itself and take the specific heat capacity of water to be 4.19 kJ/kgK (see Appendix 4).

4.3 ISOLATED SYSTEMS WITH NO PHASE CHANGE

When no energy flow takes place across the boundary of a closed system, it is an isolated system. There may be a flow of heat from hotter to colder substances until they are all at the same temperature. The study of such heat exchanges is often called **calorimetry** because, according to the old theory, 'caloric' was exchanged during the process.

Example 4.3

A steel block of mass 500 g and at a temperature of 300°C is placed in an insulated vessel containing 10 L of water at 20°C. What will be the equilibrium temperature? The specific capacity of the vessel itself may be neglected and other specific heat capacities taken from Appendix 4.

Solution

Since the temperature of the steel is above the boiling point of water, there will be some flashing of water into steam. However, since the steam cannot escape, it will eventually recondense back into water and therefore does not affect the problem. Refer to Figure 4.4.

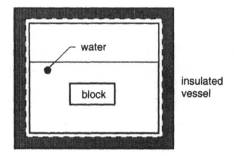

Fig. 4.4

At equilibrium, the temperature will be the same for both the water and the steel block; let this temperature be T°C.

For the steel block:

$$\text{Initial temperature } T_1 = 300°C$$
$$\text{Final temperature } T_2 = T$$
$$Q_s = mc_s(T_2 - T_1) = 0.5 \times 0.460 \times (T - 300) \text{ kJ}$$
$$= 0.23(T - 300) \text{ kJ}$$

For the water:

> Initial temperature $T_1 = 20°C$
> Final temperature $T_2 = T$
> $$Q_w = mc_w(T_2 - T_1) = 10 \times 4.19 \times (T - 20) \text{ kJ}$$
> $$= 41.9(T - 20) \text{ kJ}$$

Since no heat flows across the system boundary,

$$Q_s + Q_w = 0$$
$$\therefore 0.23(T - 300) + 41.9(T - 20) = 0$$
$$0.23T - 69 + 41.9T - 838 = 0$$
$$42.13T - 907 = 0$$
$$T = \mathbf{21.53°C}$$

Self-test problem 4.2

Two litres of hot liquid at 85°C are poured into a container of mass 0.8 kg, which is at 15°C. Determine the equilibrium temperature if no external heat transfers occur. The specific heat capacity of the liquid = 3.76 kJ/kgK, specific heat capacity of the container = 1.24 kJ/kgK, and the relative density of the liquid = 1.15.

4.4 ISOLATED SYSTEMS WITH A PHASE CHANGE

For these problems, the same principles apply as in systems with no phase change, remembering that:

- the change of phase takes place at constant temperature;
- the heat required for the phase change is $Q = mL$ (Equation 3.12);
- if equilibrium occurs before the phase change is complete, the equilibrium temperature is equal to the phase-change temperature (for example, icebergs do not melt in water at 0°C).

These principles are best illustrated by means of the next example.

Example 4.4

An insulated vessel contains 2 L of water at 15°C. A block of ice of mass 0.2 kg at –20°C is placed in the water. What is the equilibrium temperature? The latent heat of fusion of water may be taken as 335 kJ/kg and the thermal capacity of the vessel itself may be neglected. Use Appendix 4 for other specific heat capacities.

Solution

Let T be the equilibrium temperature in °C.
For the water:

> $$Q_w = m_w c_w(T_2 - T_1) = 2 \times 4.19 \times (T - 15) = 8.38(T - 15)$$

For the ice, divide the heat transfer into three stages:

Stage 1   from –20°C to 0°C:

> $$Q_i = m_i c_i(T_2 - T_1) = 0.2 \times 2.04 \times [0 - (- 20)] = 8.16 \text{ kJ}$$

Stage 2   melting:

> $$Q = m_i L = 0.2 \times 335 = 67 \text{ kJ}$$

Stage 3   from 0°C to T:

> $$Q = m_i c_w(T_2 - T_1) = 0.2 \times 4.19 \times (T - 0) = 0.84T$$

Note   Now that the ice is water, use 4.19 not 2.04 for the specific heat capacity. Total for the ice:

$$Q_i = 8.16 + 67 + 0.84T = 75.16 + 0.84T$$

Hence

$$Q_w + Q_i = 0$$
$$\therefore 8.38(T - 15) + 75.16 + 0.84T = 0$$
$$\therefore 9.22T - 50.54 = 0$$
$$T = \textbf{5.48°C}$$

Example 4.5

Repeat Example 4.4, assuming the vessel contains 1 L of water.

Solution

Q_i is the same.

$$\therefore\ Q_i = 75.16 + 0.84T$$
$$Q_w = 1 \times 4.19 \times (T - 15) = 4.19T - 62.85$$

Now $Q_i + Q_w = 0$. Therefore

$$75.16 + 0.84T + 4.19T - 62.85 = 0$$
$$5.03T + 12.31 = 0$$
$$\therefore\ T = -2.45°C$$

This answer is clearly impossible, since it means the water equilibrium temperature is below the melting point of ice. This would require all the water to refreeze, which would involve an extraction of a further 335 kJ of heat. The equation is incorrect because of the inherent assumption that *all* the ice melts. Evidently it does not; when some of the ice has melted, equilibrium occurs (with both ice and water at 0°C). Therefore

$$T = \textbf{0°C}$$

It is now possible to calculate the percentage of ice that melts.
If x is the percentage of ice that melts, then

$$Q_i = 0.2 \times 2.04 \times [0 - (-20)] + \frac{x}{100} \times 0.2 \times 335$$
$$= 8.16 + 0.67x$$
$$Q_w = 1 \times 4.19 \times (0 - 15)$$
$$= -62.85 \text{ kJ}$$
$$Q_i + Q_w = 0$$
$$\therefore 8.16 + 0.67x - 62.85 = 0$$
$$\therefore x = \textbf{81.6\%}$$

 ## Self-test problem 4.3

One kilogram of ice at –5°C is mixed with 2 kg of water at 18°C in an insulated copper vessel of mass 1.5 kg. At equilibrium, the vessel contains both ice and water. Determine the final mass of ice in the vessel. Use Appendix 4 for specific heat capacities.

4.5   BOMB CALORIMETER

The energy content or heating value of a solid or liquid substance may be determined by burning a measured quantity of the substance in an enclosed vessel (bomb) in the presence of oxygen. The bomb is immersed in a can, which contains a measured quantity of water, and by measuring the water temperature before and after firing, the heat liberated by the combustion can be determined. A very accurate thermometer fitted with an eyepiece is used so that water temperature can be measured to three decimal places.

In some designs of calorimeter, insulation is used to minimise the heat loss, and in such cases, it is necessary to plot a temperature–time diagram and apply a correction to the temperature rise to take into account the heat loss. In the **adiabatic bomb calorimeter** (Fig. 4.5), such correction is not necessary because the heat loss is virtually zero. In this adiabatic calorimeter, instead of insulation, the jacket contains water and the temperature of the water is controlled by thermistors connected to an external heating system. The jacket-water temperature is, at all times, kept the same as the temperature of the water in the can containing the bomb. Since there is no temperature difference, no heat loss occurs.

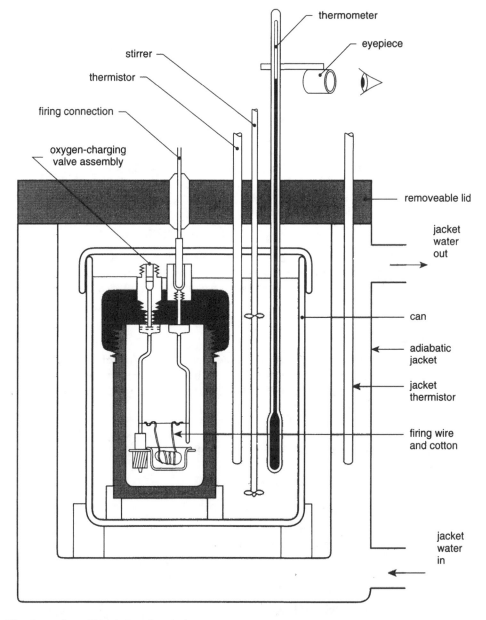

Fig. 4.5 *An adiabatic bomb calorimeter*

The bomb calorimeter is an isolated system; hence all the energy of combustion is transferred into the internal energy of the water in the can and of the metal in the bomb and the can. It is customary to account for the heat absorbed by the metal parts by first using the bomb with a substance of known and constant energy content (e.g. benzoic acid), and from the observed temperature rise of the water, the heat capacity of the metal parts can be determined. Once this has been established, it will not change and may be used for all future determinations.

Example 4.6

In a bomb calorimeter, the can is filled with 2.1 kg of water. 1.2 g of benzoic acid are burnt and the observed temperature rise is 2.94°C. If the energy content of the benzoic acid is 26.442 MJ/kg, determine the heat capacity of the metal parts of the calorimeter. The specific heat capacity of water over the temperature range is 4.187 kJ/kgK.

Solution

$$\text{Heat to water} = mc\Delta T$$
$$= 2.1 \times 4.187 \times 2.94$$
$$= 25.85 \text{ kJ}$$
$$\text{Heat of combustion} = 1.2 \times 10^{-3} \times 26.442 \times 10^3 \text{ kJ}$$
$$= 31.73 \text{ kJ}$$
$$\therefore \text{ Heat absorbed by metal parts} = 31.73 - 25.85$$
$$= 5.88 \text{ kJ}$$
$$\therefore \text{ Heat capacity of metal parts} = \frac{5.88}{2.94}$$
$$= \textbf{2.0 kJ/K}$$

Note The heat capacity of the metal parts is also often expressed as a 'water equivalent', that is, the equivalent mass of water that has the same thermal capacity as the metal parts. In this case because c water = 4.187 kJ/kgK, the water equivalent is 2/4.187 = 0.478 kg.

Example 4.7

The calorimeter given in Example 4.6 was used with a fuel of unknown energy content. The mass of fuel burnt was 0.975 g; the mass of water in the can was 2.23 kg; the temperature after firing was 23.506°C and before firing was 20.484°C. Determine the energy content of the fuel.

Solution

$$\text{Temperature rise} = 23.506 - 20.484$$
$$= 3.022°C$$
$$Q \text{ to water} = 2.23 \times 4.187 \times 3.022$$
$$= 28.216 \text{ kJ}$$
$$Q \text{ to metal} = 2.0 \times 3.022$$
$$= 6.044 \text{ kJ}$$
$$\therefore Q \text{ from fuel} = 28.216 + 6.044$$
$$= 34.26 \text{ kJ}$$
$$\therefore E = \frac{34.26}{0.975 \times 10^{-3}}$$
$$= \textbf{35.1 MJ/kg}$$

Self-test problem 4.4

A bomb calorimeter experiment resulted in the following observations:

> mass of fuel burnt = 1.02 g
> mass of water in the calorimeter = 2.52 kg
> water temperature before firing = 19.615°C
> water temperature after firing = 21.802°C
> water equivalent of the metal parts = 0.436 kg
> specific heat capacity of water (over the temperature range) = 4.175 kJ/kgK

Determine the energy content of the fuel using this data.

4.6 OPEN SYSTEMS

An open system is one in which mass transfers across the system boundary. The boundary is usually fixed, and so the volume enclosed by the boundary remains constant. Therefore, this system is also known as the **control-volume system**. Four examples of an open system are shown in Figure 4.6.

Notes

1 Since both mass and energy transfer across the system boundary, they both need to be analysed.

2 Open systems will be treated by steady-flow analysis. Steady flow means that both the mass and energy transfer rates across the system boundary at any position do not change with time. Also, any properties such as pressure or temperature measured at these positions do not change with time.

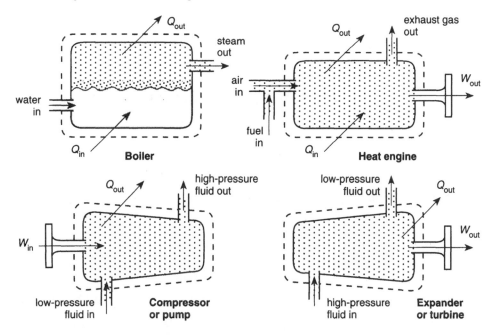

Fig. 4.6 *Open systems*

4.7 *MASS FLOW IN OPEN SYSTEMS*

If the conservation of mass principle is applied to a steady-flow open system, it follows that the amount of matter entering a system in any given time interval must be equal to the amount of matter leaving the system during the same time interval. Write the **mass-flow rate** as

$$\dot{m} = \frac{m}{t}$$

where m = mass (kg)
        t = time (s)

Steady flow implies that $\dot{m}$ is constant at any boundary of the system, and the continuity equation is

$$\boxed{\sum \dot{m}_{\text{in}} = \sum \dot{m}_{\text{out}} = \text{constant}}$$ continuity equation for mass flow (4.3)

In this equation, the Σ sign means 'sum of'. That is, the sum of the mass-flow rates into the system must equal the sum of the mass-flow rates out of the system.

Relationships between flow and velocity

Consider a time interval t during which fluid of mass (m) and volume V moves a distance x across a system boundary of cross-sectional area A (Fig 4.7).

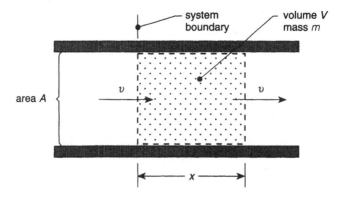

Fig. 4.7

Now $x = \upsilon t$ and $V = xA$. Therefore

$$V = \upsilon t A$$

$$\frac{V}{t} = \upsilon A$$

or $$\boxed{\dot{V} = \frac{V}{t} = \upsilon A}$$ volume-flow rate (4.4)

where $\dot{V}$ is the **volume-flow rate** and υ is the average flow velocity.

If the fluid has density ρ:

$$\rho = \frac{m}{V}$$

$$\therefore \rho = \frac{m}{vtA}$$

$$\therefore m = vtA\rho$$

or $\dfrac{m}{t} = vA\rho$

Therefore:

$$\boxed{\dot{m} = \frac{m}{t} = vA\rho}$$ mass-flow rate (4.5)

where $\dot{m}$ is the **mass-flow rate** and v is the average flow velocity.

Continuity equation with no density change

If the density of the fluid does not change appreciably between inlet and outlet, then density is a constant and the continuity equation may be written in terms of volume-flow rates thus:

$$\boxed{\sum \dot{V}_{in} = \sum \dot{V}_{out} = \text{constant}}$$ continuity equation with no density change (4.6)

The assumption of constant density is usually sufficiently accurate with liquids despite changing temperatures and pressures. However, with gases, this assumption is accurate only if there are small changes to the pressure and temperature of the gas.

Example 4.8
Water flows in a pipe of diameter 200 mm with a velocity of 5 m/s. Calculate the volume-flow rate and mass-flow rate.

Solution

$$A = \pi \times \frac{0.2^2}{4} = 0.0314 \text{ m}^2$$

From Equation 4.4

$$\dot{V} = vA = 5 \times 0.0314$$
$$= \mathbf{0.157 \text{ m}^3/s} \text{ (volume-flow rate)}$$

From Equation 4.5

$$\dot{m} = vA\rho = 5 \times 0.0314 \times 1000$$
$$= \mathbf{157 \text{ kg/s}} \text{ (mass-flow rate)}$$

Example 4.9
The inlet duct of a jet engine is 400 mm in diameter and the outlet duct 300 mm. The steady-flow velocity of air at the entrance to the inlet duct is 20 m/s and the air density

is 1.25 kg/m³. The liquid fuel supplied to the engine has relative density 0.85 and flows at a rate of 0.23 L/s. If the density of the exhaust gas is 0.35 kg/m³, calculate the exhaust-gas velocity.

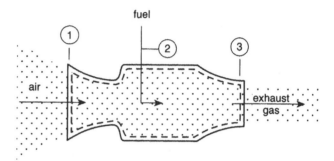

Fig. 4.8

Solution

Identify boundaries ①, ② and ③ as shown in Figure 4.8.

$$\dot{m}_1 = v_1 A_1 \rho_1 = 20 \times A(0.4) \times 1.25$$
$$= 3.142 \text{ kg/s}$$
$$\dot{m}_2 = 0.23 \times 0.85 = 0.196 \text{ kg/s}$$

From continuity:

$$\dot{m}_3 = \dot{m}_1 + \dot{m}_2 = 3.338 \text{ kg/s}$$

Now $\dot{m} = v A \rho$.

$$\therefore v_3 = \frac{\dot{m}_3}{A_3 \rho_3} = \frac{3.338}{A(0.3) \times 0.35}$$
$$= \textbf{134.9 m/s}$$

Self-test problem 4.5

In order to check piston leakage in a large internal-combustion engine, some measurements were taken as listed below:

> air-flow rate into the engine = 625 L/s and density of 1.2 kg/m³
> air/fuel ratio (mass-flow rate of air/mass-flow rate of fuel) = 14.5:1
> exhaust-gas velocity in 200 mm diameter exhaust pipe = 33.7 m/s and density of 0.75 kg/m³

Determine the leakage rate in kg/s using this data.

4.8 STEADY-FLOW ENERGY EQUATION

The general case of an *open* system is shown in Figure 4.9. Just as in a closed system, heat and work can transfer across the system boundary. However, unlike a closed system, work is involved in the flow of substance into and out of the system. This work is often called **flow work** to distinguish it from external work such as shaft work or friction work, which may also be transferred.

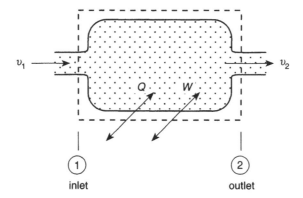

Fig. 4.9   *A general case of an open system*

Flow work may be calculated by imagining a massless, frictionless piston acting at the system boundary. Figure 4.10 shows such a model at inlet conditions.

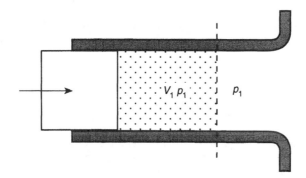

Fig. 4.10   *Model of inlet conditions*

The pressure at the inlet boundary is p_1, and the work done in moving fluid of volume V_1 across the boundary where pressure p_1 acts is

$$W = -p_1 V_1$$

(negative because work transfers *into* the system).
Similarly at the *outlet* boundary,

$$W = p_2 V_2$$

(positive because work transfers *out* of the system). Hence the change in flow work is

$$p_2 V_2 - p_1 V_1$$

The non-flow energy equation may now be modified to include flow work:

$$Q - W - (p_2 V_2 - p_1 V_1) = U_2 - U_1$$
$$\therefore Q - W = U_2 - U_1 + p_2 V_2 - p_1 V_1$$

or
$$Q - W = U_2 + p_2 V_2 - (U_1 + p_1 V_1)$$

The combination of properties $U + pV$ occurs often in thermodynamics, and is given a special name and symbol. It is called **enthalpy H**.

$$\boxed{H = U + pV}$$ enthalpy (4.7)

Hence the steady-flow energy equation becomes:

$$\boxed{Q - W = H_2 - H_1}$$ steady-flow energy equation (4.8)

Notes

1 Substitution in the steady-flow energy equation may be done in one of two ways, namely either using a convenient mass of substance (such as 1 kg) or using a convenient time interval (such as 1 s). If a time interval of 1 s is chosen, the heat and work transfer rates are in units of joules per second or watts, and hence are power.

2 The above analysis has neglected potential-energy and kinetic-energy changes. These are usually negligible because of the relatively low density of fluids (particularly gases). Some exceptions are

   - water turbines, which derive significant power from a change in potential energy;
   - high-speed water or gas flows, which may occur in nozzles, jet engines and rockets, where the kinetic energy may be of great importance.

3 Since enthalpy is a *property* of a substance, enthalpy values may be tabulated in property tables as a function of pressure and temperature. This is done by computing the specific enthalpy h, that is the enthalpy per kilogram. Hence:

$$H = mh$$

and

$$\boxed{H_2 - H_1 = m(h_2 - h_1)}$$ enthalpy change (4.9)

Also, since enthalpy is the sum of the internal energy and the flow work, the units of enthalpy are joules, and of specific enthalpy are joules per kilogram.

4 For a gas, if the specific heat capacity does not vary significantly over the range of pressure and temperature involved, the change in specific enthalpy is given by:

$$\boxed{h_2 - h_1 = c_p(T_2 - T_1)}$$ specific enthalpy change for a gas (4.10)

where c_p = specific heat capacity at constant pressure. This is discussed further in Chapter 5.

4.9 *TURBINE*

A turbine is a rotary machine in which fluid energy is converted into mechanical energy. Common types of turbines are water turbines, gas turbines and steam turbines.

Example 4.10

The power output from the steam turbine shown in Figure 4.11 is 500 kW when the steam flow rate is 1.4 kg/s.

The specific enthalpy of the steam at inlet is 3033 kJ/kg and at outlet is 2653 kJ/kg. Determine the heat loss per second from the turbine if changes in potential and kinetic energy are neglected.

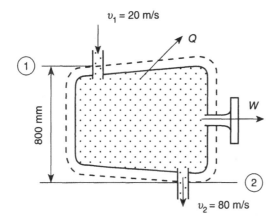

Fig. 4.11

Solution

$$Q - W = H_2 - H_1 \tag{4.8}$$
$$= m(h_2 - h_1) \tag{4.9}$$

Using a time interval of 1 s and working in kilojoule units,

$$Q - 500 = 1.4 \times (2653 - 3033) = -532 \text{ kJ}$$
$$\therefore Q = -32 \text{ kJ/s}$$
$$= \mathbf{-32 \ kW}$$

Note the negative sign, which implies heat flow *out of* the system.

4.10 *STEAM GENERATOR (BOILER)*

A steam generator or boiler (Fig. 4.12, page 82) *converts water, known as feedwater, into steam.* The necessary heat energy may be provided by combustion of a fuel (liquid, solid or gas) or from some other source (e.g. electrical, solar or nuclear).

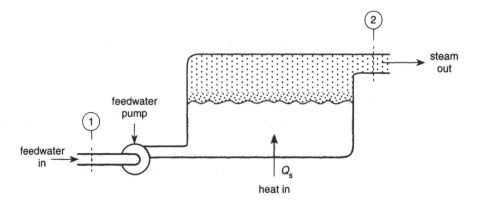

Fig. 4.12   *A steam generator*

Changes in kinetic and potential energy are negligible, so too is the work done by the feedwater pump. For example, if there is a 500 kPa increase in pressure of the feedwater going into the boiler, the work done by the feedwater pump per kg of water is: $W = pV$ $= 500 \times 1 \times 10^{-3} = 0.5$ kJ/kg. Compared with the change in specific enthalpy caused by the heat flow, this is negligible. Therefore, for a steam generator, the steady-flow energy equation reduces to:

$$Q = H_2 - H_1$$
$$= m(h_2 - h_1)$$

Example 4.11

Steam with specific enthalpy 2400 kJ/kg is produced at a rate of 2000 kg/h from feedwater with specific enthalpy 160 kJ/kg. If heat losses are 10% of the heat supplied, determine the heat supplied (Q_s) by the fuel per hour. Neglect changes to potential and kinetic energy and the work input of the feedwater pump.

Solution

$$Q = H_2 - H_1 = m(h_2 - h_1)$$
$$= 2000 \times (2400 - 160) \text{ kJ}$$
$$= 4480 \text{ MJ}$$

But because of heat losses,

$$Q_s = Q + 0.1Q_s$$
$$\therefore 0.9Q_s = 4480$$
$$Q_s = \textbf{4978 MJ/h}$$

4.11   *HEAT EXCHANGER*

A heat exchanger is a flow device in which there is an exchange of heat from a hotter substance to a colder one while keeping the substances separate.

Heat exchangers usually involve two fluids, either gas–gas, liquid–liquid or liquid–gas, but one or both substances could also be solid. In this book, the term *heat exchanger* will always be used with *fluids in flow*; hence, the heat exchanger performs an equivalent function to its non-flow counterpart, the calorimeter.

Figure 4.13 illustrates a **crossflow heat exchanger**, so named because the fluid flow paths form a cross. Heat exchangers can also have a **parallel-flow, counterflow** or **mixed-flow** configuration.

If there is no transfer of work, the steady-flow energy equation applied to either fluid is

$$Q = H_2 - H_1$$

where the subscripts 1 and 2 refer to inlet and outlet boundaries respectively, as before.

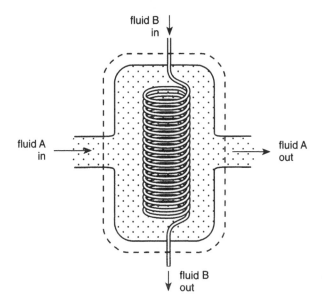

Fig. 4.13 *A crossflow heat exchanger*

Heat exchangers are generally insulated or constructed so that external heat exchanges are negligible. Hence

$$H_2 - H_1 + H_2 - H_1 = 0$$
(fluid A)   (fluid B)

or

$$H_2 - H_1 = H_1 - H_2$$
(fluid A)   (fluid B)

That is, the enthalpy gain by one fluid equals the enthalpy loss from the other. This relationship is valid for all types of heat exchangers and all types of fluids, provided external heat losses or gains are negligible.

The enthalpy data for the fluids may be obtained from tables, given the appropriate pressures and temperatures, and using

$$H_2 - H_1 = m(h_2 - h_1)$$

Note   The heat capacity of the metal parts of the exchanger is irrelevant if steady-flow conditions exist; only the heat transfers between the fluids need to be considered.

If no phase change occurs in a fluid and the specific heat capacity is approximately constant over the temperature range, then

$$H_2 - H_1 = mc(T_2 - T_1)$$

For gases, the specific heat capacity at constant pressure (c_p) should be used in this equation.

Example 4.12

A fresh-water heat exchanger is used to cool the oil in a marine engine. Water temperature into the exchanger is 15°C and water temperature out is 30°C. Oil temperature in is 150°C and oil flow is 80 L/min. The oil has a relative density of 0.9 and a specific heat capacity of 1.8 kJ/kgK. The specific heat capacity of water may be taken as 4.19 kJ/kgK.

If the outlet oil temperature is 100°C, determine the flow rate of water.

Solution

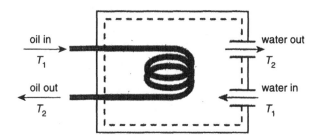

oil in
T_1

oil out
T_2

water out
T_2

water in
T_1

Fig. 4.14

Use a time interval of 1 s.
For the water:

$$Q = mc(T_2 - T_1)$$
$$= m \times 4.19 \times (30 - 15)$$
$$= 62.85m$$

For the oil:

$$Q = \frac{80}{60} \times 0.9 \times 1.8 \times (150 - 100)$$
$$= 108 \text{ kJ/s}$$
$$\therefore 62.85m = 108$$
$$\therefore m = 1.72 \text{ kg/s (L/s)}$$

 ## Self-test problem 4.6

A water-cooled condenser for steam condenses 25 kg/s of steam. The specific enthalpy of the steam into the condenser is 2838 kJ/kg and the specific enthalpy of the condensed steam (condensate) is 209 kJ/kg. Determine:
(a) the rate at which heat is being removed from the steam in the condenser;
(b) the flow rate of cooling water if the temperature increase of the cooling water is 24°C.

4.12 *GAS CALORIMETER*

The gas calorimeter (Fig 4.15) *is used to determine the energy content of a gas.*

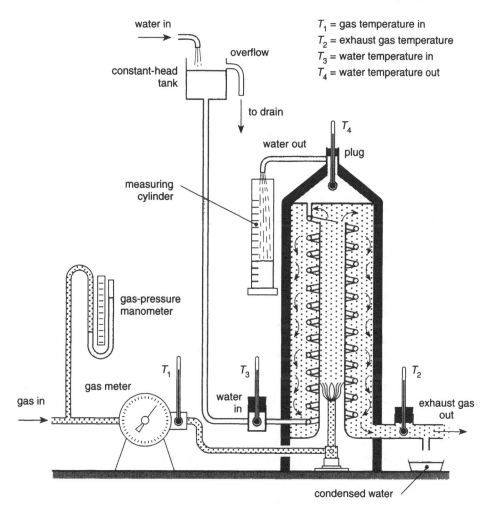

Fig. 4.15 *A gas calorimeter*

The gas passes through a constant-pressure regulator and then through an accurate gas meter before being mixed with air and burnt at constant pressure. The hot combustion products pass over a number of coils through which water circulates. The water enters and leaves the calorimeter at a constant level (head) so that the set flow rate remains constant. This flow rate may be determined by diverting the water that is leaving the calorimeter to a measuring cylinder, and measuring the amount collected in a given time.

The gas calorimeter is therefore the flow counterpart of the bomb calorimeter (non-flow): in the bomb calorimeter, combustion takes place in a closed system under constant-volume conditions, whereas in the gas calorimeter, combustion takes place in an open system under constant-pressure conditions. Also, in the gas calorimeter it is not necessary to compensate for the metal parts because readings are not taken until steady-state conditions are reached.

The heat loss from the calorimeter walls and in the exhaust gas is usually negligible if the flow rate of water is set so that the gas temperature out of the calorimeter is only a little above ambient temperature. Under these conditions, the gas calorimeter is essentially a heat exchanger where the heat of combustion is transferred to an enthalpy increase of the water.

Self-test problem 4.7

A five-minute test with a gas calorimeter yielded the following data:

> gas-meter reading: start = 2.85 L; finish = 6.52 L
> water temperature: in = 20.5°C; out = 35.3°C
> volume of water collected = 1.95 L

Determine the energy content of a gas sample in MJ/m^3 using a specific heat capacity of water = 4.187 kJ/kgK.

Summary

The two main types of systems encountered in thermodynamics are closed systems and open systems. In closed systems no mass transfers over the system boundary but energy in the form of heat or work may transfer across the boundary. The non-flow energy equation applies to such systems, namely $Q - W = U_2 - U_1$, that is, the nett energy transfer causes a change in internal energy of the substance in the system.

In open systems, mass as well as energy in the form of heat and work transfer over the system boundary both into and out of the system. The system is assumed to be in a steady-state condition, which means that mass and energy transfers are constant with time and the properties of the substance flowing through the system at any point in it do not change with time. The continuity equation applies to the mass transfers in such systems, and this is basically the conservation of mass principle applied to mass-flow rates through the system. The steady-flow energy equation applies to the energy transfers in such systems. In the most general case, potential and kinetic energy changes may occur, but in the majority of thermodynamic applications in engineering, these changes are small compared with other energy changes and can be neglected. With these assumptions the steady-flow energy equation is: $Q - W = H_2 - H_1$, that is, the nett energy transfer causes a change in enthalpy of the substance flowing through the system.

Problems

Note For these problems, unless otherwise stated, assume constant specific heat capacities taken from Appendix 4. For water, latent heat of evaporation = 2257 kJ/kg and latent heat of fusion = 335 kJ/kg.

4.1 During a non-flow process there is a decrease of 200 kJ in the internal energy of a system whereas 1 MJ of work is transferred out of the system. Determine the heat flow and state whether the heat enters or leaves the system.

> 800 kJ into the system

4.2 In a closed system containing 1.5 kg of water at 20°C, heat and work are transferred. The heat transfer is 150 kJ in, and the final temperature of the water is 48°C. Determine:
(a) the change in internal energy of the water;
(b) the work transfer (stating direction);

(c) the additional heat flow required for the same temperature increase if there were no work transfer.

(a) 176 kJ   (b) –26 kJ (in)   (c) 26 kJ

4.3 A tank for an outboard motor test contains 200 L of water at 18°C. The motor delivers a propeller power of 35 kW for 10 min on test. What will be the temperature of the water after the test if external heat exchanges from the water are negligible?

43.1°C

4.4 An aluminium saucepan of mass 0.6 kg contains 1.5 L of water at 15°C. Calculate how long it will take for the water to boil away if the saucepan is placed on a 1.2 kW electric hotplate. Take heat losses to be 40% of the hotplate heat output.

91.8 min

4.5 A cast-iron block of mass 2 kg at a temperature of 420°C is placed in an insulated vessel containing 100 L of oil at 18°C. What is the equilibrium temperature? (Assume c for oil = 2.0 kJ/kgK and RD oil = 0.75.)

20.2°C

4.6 A solid of mass 0.5 kg and temperature 100°C is dropped into a vessel containing 0.5 kg of water at 20°C; the vessel is then sealed. The water temperature rises to 35°C. If external heat transfers from the water are negligible, determine the specific heat capacity of the solid.

967 J/kgK

4.7 In a bomb-calorimeter experiment, the can is filled with 2 kg of water. One gram of substance of known energy content 28.5 MJ/kg is burnt and the observed temperature rise is 2.76°C. Determine the heat capacity of the metal parts of the calorimeter. The specific heat capacity of water over the temperature range is 4.187 kJ/kgK.

1.952 kJ/K

4.8 The bomb calorimeter in Problem 4.7 is tested with a fuel of unknown energy content, and the following observations recorded:

    mass of fuel burnt = 1.25 g
    mass of water in the calorimeter = 2.75 kg
    water temperature before firing = 18.5°C
    water temperature after firing = 21.8°C

Determine the energy content of the fuel.

35.55 MJ/kg

4.9 A portable cooler consists of a sheet-steel liner of mass 0.75 kg insulated on the outside. The cooler holds 1.5 L of water at 17.5°C. A block of ice of mass 0.18 kg and at –5°C is placed in the cooler. Calculate the equilibrium temperature, neglecting external heat exchanges.

7.3°C

4.10 Repeat Problem 4.9 with a mass of ice of 0.38 kg. Calculate the percentage of ice that melts.

88%

4.11 Two liquids, A and B (whose properties are given in the table), are mixed together in an insulated vessel in the ratio B:A = 2:1 by volume. Determine the equilibrium temperature.

	A	B
RD	0.8	0.75
c(kJ/kgK)	2.4	1.8
T(°C)	30.0	50.00

    41.7°C

4.12 Oil of relative density 0.85 flows with a velocity of 3.5 m/s through a pipe of internal diameter 300 mm. Determine the volume-flow rate and mass-flow rate of the oil.
    0.247 m³/s; 210 kg/s

4.13 Oil of relative density 0.85 flows through a pipe that connects to a single-pass heat exchanger consisting of 30 copper tubes of internal diameter 25 mm. If the velocity of the oil in the tubes and the pipe is 3 m/s, determine:
(a) the volume-flow rate;
(b) the mass-flow rate;
(c) the internal diameter of the pipe.
    (a) 44.2 L/s   (b) 37.6 kg/s   (c) 137 mm

4.14 Water flows with a velocity of 5 m/s in a pipe of diameter 300 mm. Further downstream, a pipe of 100 mm diameter connects into the 300 mm pipe. The water leaves the 100 mm pipe with a velocity of 8 m/s. Calculate:
(a) the volume-flow rate;
(b) the mass-flow rate;
(c) the velocity in the 300 mm pipe downstream of the junction.
    (a) 0.353 m³/s   (b) 353 kg/s   (c) 4.11 m/s

4.15 A large internal-combustion engine takes in air at 2 kg/s and fuel of relative density 0.85 at 8.8 L/min. The exhaust pipe has a diameter of 400 mm and the exhaust gas has a density of 0.4 kg/m³. Assuming steady-flow conditions, calculate the velocity of the exhaust gas.
    42.3 m/s

4.16 A single-acting, single-cylinder air compressor has a bore of 150 mm and a stroke of 180 mm and runs at 720 rpm. Calculate the theoretical (loss-free) mass-flow rate of air through the compressor if the inlet air has a density of 1.25 kg/m³.
    If the outlet air has a density of 6 kg/m³, calculate the theoretical inlet-pipe and outlet-pipe diameters for an air velocity of 15 m/s in both pipes.
    0.0477 kg/s; 57 mm; 26 mm

4.17 A coal-fired steam generator under steady-flow conditions uses coal, with non-combustible content 5%, at a rate of 1.8 t/h. The air–fuel ratio (by mass) is 18:1. The exhaust gas enters a stack of diameter 1.8 m at which point the gas has a specific volume of 3.6 m³/kg. Assuming the exhaust gas is free of fly-ash, determine:
(a) the volume-flow rate and mass-flow rate of exhaust gas;
(b) the velocity of gas entering the stack.
    (a) 34.11 m³/s; 9.475 kg/s   (b) 13.4 m/s

4.18 In a marine diesel-engine installation, some of the energy in the exhaust gas is used to provide hot water for the ship. The hot-water storage tank has capacity 12 m³ and is heated to 85°C from cold (10°C) in 4 h. If the flow rate of exhaust gas is 8.5 kg/s, determine the temperature decrease in the exhaust gas due to hot-water heating. Assume the specific heat capacity (at constant pressure) for the exhaust gas is 1.08 kJ/kgK.

    28.5°C

4.19 A steam turbine takes in steam, with specific enthalpy 3095 kJ/kg, at a rate of 80 kg/min. The steam leaves the turbine with specific enthalpy 2660 kJ/kg. If heat losses from the turbine are 120 kW, determine the power output. Assume that kinetic-energy and potential-energy changes are negligible.

    460 kW

4.20 In a steady-flow system, the mass-flow rate is 12 kg/min and the nett heat flow is 85 kW into the system. The gain in specific enthalpy of the working substance is 200 kJ/kg. Neglecting potential-energy and kinetic-energy changes, calculate the power output from the system.

    45 kW

4.21 Refrigerant flows through a water-cooled condenser at a rate of 25 kg/min. The specific enthalpy of refrigerant entering the condenser is 400 kJ/kg, and leaving, is 220 kJ/kg. Determine the mass flow of cooling water through the condenser for a temperature increase of the water of 10°C, assuming no external heat exchanges.

    1.79 kg/s

4.22 Air enters the radiator of a motor vehicle at 15°C and at a rate of 240 m³/min, and leaves the radiator at a temperature of 25°C. If water circulates through the radiator at a rate of 50 L/min, calculate the temperature drop of the water. The specific volume of the air entering the radiator is 0.8 m³/kg.

    14.4°C

4.23 After steady-state conditions were achieved in a gas calorimeter, the inlet temperature of the cooling water was 16°C and the outlet temperature was 21°C. The flow rate of water was 2.3 L/min and the flow rate of gas was 2.8 L/min. Determine the energy content of the gas. Use specific heat capacity of water = 4.187 kJ/kgK.

    17.2 MJ/m³

4.24 A steam generator uses 3 t of fuel per hour in producing steam with a specific enthalpy of 2800 kJ/kg from feedwater with a specific enthalpy of 120 kJ/kg. The energy content of the fuel is 28.5 MJ/kg. If heat losses are 30%, determine the mass-flow rate of steam. Neglect changes in kinetic and potential energy.

    6.2 kg/s

CHAPTER 5

Gases

≡

Objectives

On completion of this chapter you should be able to:

- define a perfect or ideal gas in terms of the molecular model and explain when real gases approach ideal gas behaviour;
- calculate property changes to a gas undergoing typical thermodynamic processes such as heating, cooling, compression or expansion;
- describe the following gas processes: constant pressure, constant volume, isothermal, polytropic and adiabatic (isentropic). For each of these processes, draw a p–V diagram and calculate property changes or hear or work transfers.

Introduction

This chapter deals with the laws and processes applicable to gases and mixtures of gases. The term *gas* will be used henceforth in the general sense and will also mean gas mixtures.

First, a perfect or ideal gas is defined, then the general equations valid for *any* process on a fixed mass of ideal gas are treated. Heating of a gas and the specific heat capacity of a gas are then investigated. Particular common types of processes applicable to gases are then treated in turn, namely constant pressure, constant volume, isothermal, polytropic and adiabatic (isentropic). The chapter concludes with a comparison and summary of the various gas processes.

In all cases, it is assumed that there is a constant mass of gas, that is, the gas is contained in a non-flow or control-mass system. The analysis is generally not valid for a flow system, that is, a control-volume system, where the gas is flowing through the system. Perfect or ideal gas behaviour is also assumed throughout.

5.1 PERFECT OR IDEAL GAS

In this chapter, gases are treated as it they are **perfect** or **ideal**. A perfect or ideal gas is sometimes defined as one that obeys the gas laws but this is not a very illuminating definition. In terms of molecular theory, a perfect gas is one in which:

- the molecules of the gas occupy negligible volume compared with the volume of the containing vessel;
- the molecules move independently of one another except when a collision occurs. A collision causes an internal redistribution of energy between the colliding molecules;
- energy transfers into or out of the gas cause a change in molecular linear kinetic energy only; hence any nett energy transfers result in a proportional change in temperature. This means that the specific heat capacity of a perfect gas is constant and does not change with temperature or pressure.

The assumption of perfect gas behaviour is usually a good approximation for many real gases such as:

- monoatomic gases such as helium and argon;
- diatomic gases such as nitrogen, oxygen and carbon dioxide;
- gas mixtures such as air.

It is not such a good assumption for vapours such as refrigerants and steam, whose properties are usually obtained from property charts or tables.

Perfect gas behaviour is not a good assumption for a real gas if:

- the gas is close to liquefaction, either highly compressed and/or at a low temperature, because in this case, the molecules are close together;
- the gas is at a very high temperature, because dissociation and electron excitation may occur, which involves energy other than linear kinetic energy.

5.2 *GENERAL GAS EQUATIONS*

Several equations are valid for *any* process on a fixed mass of perfect gas. These might be termed general equations. They are:

$$\frac{pV}{T} = c \quad \text{or} \quad \frac{p_1V_1}{T_1} = \frac{p_2V_2}{T_2}$$ general gas equation (5.1)

$$pV = mRT$$ equation of state (5.2)

$$R = \frac{8314}{M}$$ gas constant (5.3)

where
p = absolute pressure (Pa)
V = volume (m^3)
T = absolute temperature (K)
c = a constant
m = mass of gas (kg)
M = relative molecular mass (no units)
R = gas constant (J/kgK)

The subscripts 1 and 2 refer to initial and final conditions respectively. Because there is a fixed mass of gas, the non-flow energy equation

$$Q - W = U_2 - U_1 \tag{4.1}$$

applies in all cases.

Notes

1 The relative molecular mass M is obtained from the relative atomic mass (given in Appendix 3) of each of the atoms in a molecule of the gas, knowing the chemical formula for the gas.

2 Equations 5.1 and 5.2 are also valid for a gas mixture, provided the value of R is for the gas mixture; that is, for a gas mixture, $R = \dfrac{8314}{m}$ where M is the *mean* molecular mass of the gas mixture obtained from the ratio of the *volumes* of the constituents in the gas mixture.

3 Equation 5.2 may also be written:

$$pv = RT \quad \text{or} \quad p = \rho RT$$

where $\rho = \dfrac{m}{V}$ (density)

$\quad\quad\;\; v = \dfrac{V}{m}$ (specific volume).

4 The number 8314 is also known as the universal gas constant.

5 The gas constant R for air is 287 J/kgK. This value will be used throughout this chapter.

6 The density of air at normal atmospheric conditions is about 1.2 kg/m³. Compare this with the density of water, which is 1000 kg/m³.

Example 5.1

A mass of gas has volume 4 m³ when the temperature is 120°C and the pressure is 25 kPa (gauge). Determine the volume of this mass of gas at normal atmospheric conditions (20°C and 101.3 kPa).

Solution

$$p_1 = 126.3 \text{ kPa} \quad (25 + 101.3)$$
$$V_1 = 4 \text{ m}^3$$
$$T_1 = 393 \text{ K} \quad (120 + 273)$$
$$p_2 = 101.3 \text{ kPa}$$
$$T_2 = 293 \text{ K} \quad (20 + 273)$$

$$\frac{p_1 V_1}{T_1} = \frac{p_2 V_2}{T_2}$$

$$\therefore V_2 = \frac{p_1}{p_2} \times \frac{T_2}{T_1} \times V_1$$

$$= \frac{126.3 \times 10^3}{101.3 \times 10^3} \times \frac{293}{393} \times 4$$

$$= \mathbf{3.72 \text{ m}^3}$$

Example 5.2

Determine the value of the gas constant for hydrogen, oxygen and argon.

Solution

For hydrogen H_2:
$$M = 2 \quad (1 + 1) \quad (Appendix\ 3)$$
$$\therefore R = \frac{8314}{2}$$
$$= \textbf{4157 J/kgK}$$

For oxygen O_2:
$$M = 32 \quad (16 + 16)$$
$$\therefore R = \frac{8314}{32}$$
$$= \textbf{259.8 J/kgK}$$

For argon Ar:
$$M = 40$$
$$\therefore R = \frac{8314}{40}$$
$$= \textbf{207.85 J/kgK}$$

Example 5.3

A compressed-air tank is a cylinder 300 mm in diameter and 2.5 m long. After filling, a pressure gauge fitted to the tank reads 1 MPa when atmospheric pressure is 101.3 kPa and the temperature of the air in the tank is is 20°C. Calculate the mass of air in the tank.

Solution

Tank volume:
$$V = A(0.3) \times 2.5$$
$$= 0.1767 \text{ m}^3$$

Absolute pressure:
$$p = 1 \times 10^6 + 101.3 \times 10^3$$
$$= 1.1013 \times 10^6 \text{ Pa}$$
$$R \text{ for air} = 287 \text{ J/kgK} \quad (\text{see note 5})$$
$$pV = mRT$$
$$\therefore m = \frac{pV}{RT}$$
$$= \frac{1.1013 \times 10^6 \times 0.1767}{287 \times 293}$$
$$= \textbf{2.314 kg}$$

Absolute temperature:
$$T = 20 + 273$$
$$= 293 \text{ K}$$

Self-test problem 5.1

A valve is opened in the tank given in Example 5.3 and air escapes until equilibrium occurs. Determine the mass of air that escapes.

5.3 SPECIFIC HEAT CAPACITY OF A GAS

Heat is not a property of a substance and the quantity of heat transferred (for a given temperature rise) depends on the method or process by which the heat is transferred. Solids or liquids are almost always heated at constant pressure.

Some of the common ways in which **constant-pressure heating** takes place are illustrated in Figure 5.1.

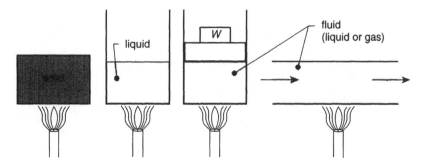

Fig. 5.1   *Constant-pressure heating*

With gases, another common method is **constant-volume heating**. This occurs when the gas is confined in a rigid vessel such that expansion in volume is negligible. This is illustrated in Figure 5.2.

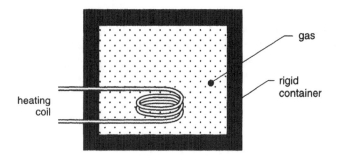

Fig. 5.2   *Constant-volume heating*

The amount of heat necessary to give the same temperature rise is always greater for constant-pressure heating than for constant-volume heating, that is, *the specific heat capacity at constant pressure (c_p) is greater than the specific heat capacity at constant volume (c_v).* This is seen from an analysis of Figure 5.3 (page 95), which shows that in the case of constant-pressure heating, more energy is necessary (for the same temperature rise) because work is also done.

Constant-pressure and constant-volume heating are just two of the possible ways a gas may be heated. Other methods are dealt with later on in this chapter.

The amount of heat necessary to cause a temperature change from T_1 to T_2 when a gas is heated is given by:

$$Q = mc_p(T_2 - T_1)$$ constant-pressure heating (5.4)

$$Q = mc_v(T_2 - T_1)$$ constant-volume heating (5.5)

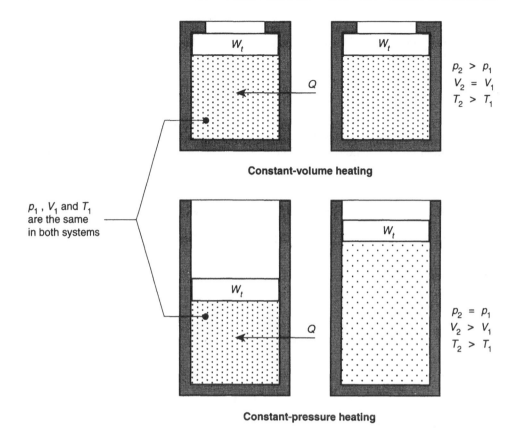

Constant-volume heating

p_1, V_1 and T_1
are the same
in both systems

Constant-pressure heating

Fig. 5.3 *More heat is needed to obtain the same temperature increase with constant-pressure heating compared with constant-volume heating*

Notes

1 The value of c_p and c_v for air are given in Appendix 4 and are 1005 J/kgK and 718 J/kgK respectively.

2 The gas constant R is related to specific heat capacity at constant pressure and volume by the following equation (derived in Appendix 6):

$$R = c_p - c_v$$ gas constant (5.6)

For example, R for air is 287 J/kgK. Now c_p for air is 1005 J/kgK and c_v is 718 J/kgK. It is indeed evident that $R = c_p - c_v$.

5.4 *INTERNAL ENERGY AND ENTHALPY CHANGE IN A GAS*

When a gas is heated at constant volume, no work is done, so that all the heat transfers into internal energy. Hence:

$$U_2 - U_1 = mc_v(T_2 - T_1)$$ internal energy change (5.7)

When a gas is heated at constant pressure, work is done. The heat transfer is equal to the change in enthalpy, that is, the change in internal energy plus the work $p(V_2 - V_1)$. Hence:

$$H_2 - H_1 = mc_p(T_2 - T_1)$$ enthalpy change (5.8)

Although Equations 5.7 and 5.8 were derived for constant-volume and constant-pressure heating processes respectively, they apply to *any* gas process because all the terms in these equations are properties and property relationships do not depend on the process.

5.5 *CONSTANT-PRESSURE PROCESS*

In a constant-pressure process, the gas pressure does not change.

p–V diagram

See Figure 5.4.

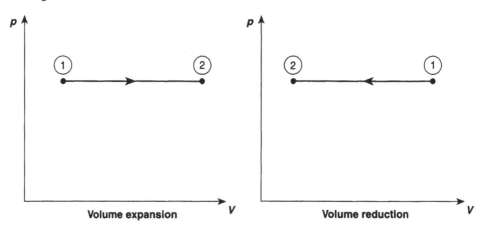

Fig. 5.4 *p–V diagrams for constant-pressure processes*

Example of the process

An example of gas compression or expansion against a constant resisting force is shown in Figure 5.5.

Relationship between *p*, *V* and *T*

Since $p_1 = p_2 = p$, the general equation becomes:

$$\frac{V_1}{T_1} = \frac{V_2}{T_2}$$ constant-pressure process (5.9)

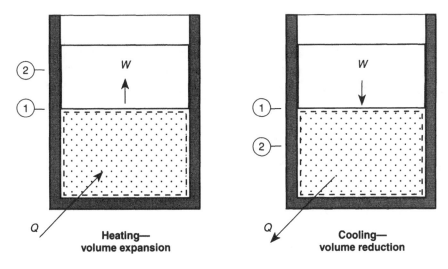

Fig. 5.5 *Examples of constant-pressure processes*

Work

It was shown in Chapter 3 that for a constant-pressure process,

$$\boxed{W = p(V_2 - V_1)}$$ constant-pressure work (3.6)

This equation (like all equations in this chapter) is valid for both volume-expansion and volume-reduction processes, remembering that V_1 is the initial volume and V_2 is the final volume. So for an expansion process, W is positive (work out of the system), and for a reduction process, W is negative (work into the system).

Heat

For a constant-pressure process, the specific heat capacity at constant pressure (c_p) must be used. Equation 5.4 applies:

$$\boxed{Q = mc_p\,(T_2 - T_1)}$$ (5.4)

Again, the sign of the heat flow is correct, remembering that T_1 is the initial temperature and T_2 the final temperature.

Also for a constant-pressure heating process,

$$Q = H_2 - H_1$$

and

$$H_2 - H_1 = mc_p\,(T_2 - T_1)$$

Example 5.4

A piston-and-cylinder mechanism, as illustrated in Figure 5.6, contains air and is in equilibrium with the surrounding air, which is at 15°C. Heat energy is supplied and the piston rises 400 mm. Neglect friction and assume the following values for air:
$c_p = 1005$ J/kgK, $c_v = 718$ J/kgK, $R = 287$ J/kgK. Atmospheric pressure is 101.3 kPa.
Determine:
(a) the mass of air;
(b) the final temperature;
(c) the heat flow;
(d) the internal-energy change;
(e) the work transfer.
Also, verify that the non-flow energy equation balances, and account for the work transfer as an energy change in the surroundings.

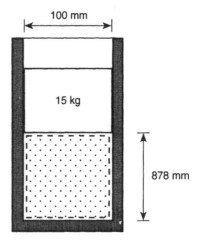

Fig. 5.6

Solution

(a) The cross-sectional area is $A(0.1) = 7.854 \times 10^{-3}$ m²

$$V_1 = 7.854 \times 10^{-3} \times 0.878 \text{ m}^3$$
$$= 6.896 \text{ L}$$

The pressure in the system is atmospheric plus the pressure due to the piston weight.

$$\therefore p_1 = 101.3 \times 10^3 + \frac{15 \times 9.81}{7.854 \times 10^{-3}} \text{ Pa}$$
$$= 120 \text{ kPa}$$
$$T_1 = 288 \text{ K } (15°C)$$

From Equation 5.2:

$$p_1V_1 = mRT$$
$$\therefore m = \frac{p_1V_1}{RT_1}$$

Substituting:

$$m = \frac{120 \times 10^3 \times 6.896 \times 10^{-3}}{287 \times 288}$$

$$= \mathbf{10.01 \times 10^{-3}\ kg}$$

(b) $V_2 = 7.854 \times 10^{-3} \times (0.878 + 0.4)\ \text{m}^3$

$$= 10.04\ \text{L} \quad \text{(actually 10.0374 L, retain in calculator memory)}$$

Since this is a constant-pressure process,

$$\frac{V_1}{T_1} = \frac{V_2}{T_2}$$

$$\therefore T_2 = \frac{V_2 \times T_1}{V_1}$$

$$= \frac{10.04 \times 10^{-3} \times 288}{6.896 \times 10^{-3}}$$

$$= \mathbf{419.2\ K\ (146.2°C)}$$

(c) From Equation 5.4:

$$Q = mc_p (T_2 - T_1)$$

$$= 10.01 \times 10^{-3} \times 1005 \times (419.2 - 288)$$

$$= \mathbf{1320\ J} \quad \text{(heat in)}$$

(d) From Equation 5.7:

$$U_2 - U_1 = mc_v (T_2 - T_1)$$

$$= 10.01 \times 10^{-3} \times 718 \times (419.2 - 288)$$

$$= \mathbf{943\ J} \quad \text{(increase in internal energy)}$$

(e) From Equation 3.6:

$$W = p(V_2 - V_1)$$

$$= 120 \times 10^3 \times (10.04 - 6.896) \times 10^{-3}$$

$$= \mathbf{377\ J} \quad \text{(work out)}$$

The non-flow energy equation is:

$$Q - W = U_2 - U_1$$

$$1320 - 377 = 943$$

which **balances**.

The work transfer out of the system is accounted for as:

(i) an increase in potential energy of the piston:

$$PE = mgh$$

$$= 15 \times 9.81 \times 0.4$$

$$= 58.86\ \text{J}$$

(ii) work done on the air surrounding the piston, which exerts a pressure and therefore a force to oppose the motion of the piston:

$$W = p(V_2 - V_1)$$

$$= 101.3 \times 10^3 \times (10.04 - 6.896) \times 10^{-3}$$

$$= 318.2\ \text{J}$$

Check total:

$$(i) + (ii) = 58.86 + 318.2$$
$$= \textbf{377 J}$$

which equals the work leaving the system.

 ## *Self-test problem 5.2*

A gas of mass 0.006 kg is heated at a constant pressure of 200 kPa (absolute) from a temperature of 100°C and volume 3 L to a temperature of 300°C and volume 4.6 L. The gas has a specific heat capacity of 700 J/kgK at constant volume. Determine:

(a) the internal-energy change of the gas;

(b) the work done during the expansion;

(c) the heat supplied;

(d) the specific heat capacity at constant pressure;

(e) the enthalpy change.

5.6   *CONSTANT-VOLUME PROCESS*

In a constant-volume process, the gas volume does not change.

p–V diagram

See Figure 5.7.

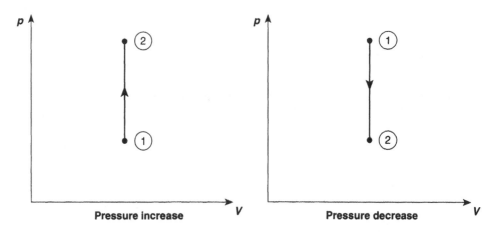

Fig. 5.7   *p–V diagrams for constant-volume processes*

Example of the process

An example of heating or cooling a mass of gas confined in a rigid container is shown in Figure 5.8.

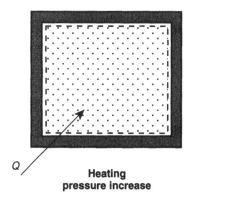 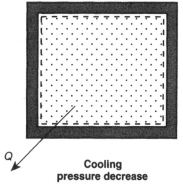

Fig. 5.8 *Examples of constant-volume processes*

Relationship between *p*, *V* and *T*

Since $V_1 = V_2 = V$, the general equation becomes:

$$\frac{p_1}{T_1} = \frac{p_2}{T_2}$$

constant-volume process (5.10)

Work

Since there is no change in volume, there is no work. This can also be seen from the p–V diagram, since the area under the vertical line = 0.

Heat

The specific heat capacity at constant volume must be used. Equation 5.5 applies:

$$Q = mc_v(T_2 - T_1)$$

(5.5)

Non-flow energy equation

Since $W = 0$, the non-flow energy equation reduces to

$$Q = U_2 - U_1$$

that is, the heat transfer equals the internal-energy change. This is also evident from the fact that $Q = mc_v(T_2 - T_1)$ and $U_2 - U_1 = mc_v(T_2 - T_1)$.

Example 5.5

Suppose the piston in Example 5.4 (Fig. 5.6) was locked in the equilibrium position (at a height of 878 mm above the bottom of the cylinder). The same amount of heat energy (1.3 kJ) is now supplied. Calculate:
(a) the final temperature;
(b) the final pressure;
(c) the internal-energy change.
Verify that the non-flow energy equation balances.

Solution

Mass of gas in the system $= 10.01 \times 10^{-3}$ kg
Initial temperature $T_1 = 288$ K
Heat flow $Q = 1320$ J
Initial volume $V_1 = 6.896$ L
Initial pressure $p_1 = 120$ kPa
$c_v = 718$ J/kgK

(a) The final temperature can be obtained from Equation 5.5:

$$Q = mc_v(T_2 - T_1)$$

$$\therefore T_2 - T_1 = \frac{Q}{mc_v}$$

$$= \frac{1320}{10.01 \times 10^{-3} \times 718}$$

$$= 183.7 \text{ K}$$

$$\therefore T_2 = 183.7 + 288$$

$$= \mathbf{471.7 \text{ K}}$$

which is higher than T_2 for the constant-pressure process, as expected.

(b) The final pressure can be obtained from Equation 5.10:

$$\frac{p_1}{T_1} = \frac{p_2}{T_2}$$

$$\therefore p_2 = \frac{T_2}{T_1} \times p_1$$

$$= \frac{471.7}{288} \times 120$$

$$= \mathbf{196.5 \text{ kPa}}$$

(c) The change in internal energy can be obtained from Equation 5.7:

$$U_2 - U_1 = mc_v(T_2 - T_1)$$
$$= 10.01 \times 10^{-3} \times 718 \times (471.7 - 288)$$
$$= \mathbf{1320 \text{ J}}$$

The non-flow energy equation is

$$Q = U_2 - U_1$$
$$1320 = 1320$$

which **balances**.

 ## Self-test problem 5.3

A motor-vehicle tyre has a volume of 35 L. It is filled with air at a temperature of 27°C until the gauge reads 280 kPa. Using the values for air given in Example 5.4:
(a) calculate the mass of air in the tyre.
The air temperature drops to 15°C with negligible change in volume.
(b) What pressure would the gauge now show?
(c) What is the heat flow?
(d) What is the internal energy change?
(e) What mass of air at 15°C would need to be added to the tyre to restore the gauge pressure to 280 kPa?

5.7   *ISOTHERMAL PROCESS*

In an isothermal process the gas temperature does not change.

p–V diagram

Since $\dfrac{pV}{T}$ = constant, then when T is constant, pV is also constant. The equation $pV = c$ where c is a constant is the equation of a rectangular hyperbola, as illustrated in Figure 5.9.

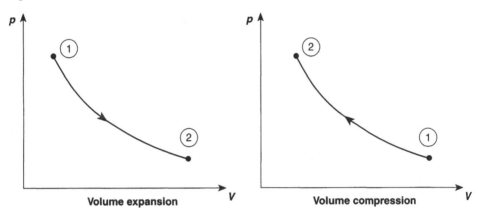

Fig. 5.9 *p–V diagrams for isothermal processes*

Example of the process

If gas in a cylinder made of heat-conducting material is slowly compressed to a higher pressure, the cylinder can be cooled so that the heat generated by compression flows out and the temperature of the gas remains constant. Conversely, if a compressed gas is slowly expanded and does work while heat flows in from the surroundings, expansion is isothermal if the temperature remains constant. These are illustrated in Figure 5.10.

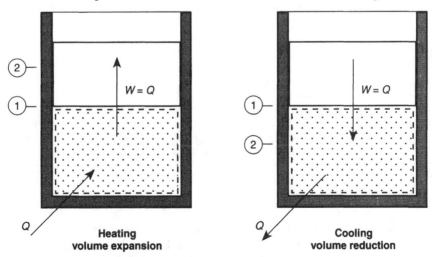

Fig. 5.10 *Examples of isothermal processes*

It can be reasoned that if the temperature remains constant, there can be no heat flow, since heat flow depends on a temperature difference. This reasoning is fallacious; the system temperature is constant but must be *different from the temperature of the surroundings* in order for the process to be isothermal. The heat flow takes place between the system and the surroundings. Therefore, in order for an isothermal process to occur, the surroundings must be at a different temperature from the gas in the system.

Relationship between p, V and T

Since $T_1 = T_2 = T$, the general equation $\dfrac{p_1 V_1}{T_1} = \dfrac{p_2 V_2}{T_2}$ reduces to:

$$\boxed{p_1 V_1 = p_2 V_2}$$ isothermal process (5.11)

Work

The work done in an isothermal process is given by the following equation (derived in Appendix 7):

$$\boxed{W = p_1 V_1 \ln \left(\frac{V_2}{V_1}\right)}$$ isothermal work (5.12)

Since $p_1 V_1 = p_2 V_2$, this equation can also be written

$$W = p_2 V_2 \ln \left(\frac{V_2}{V_1}\right)$$

Again, the sign of the work transfer is correct: for an *expansion*, $V_2/V_1 > 1$ and $\ln (V_2/V_1)$ is positive; for a *compression* process, $V_2/V_1 < 1$ and $\ln (V_2/V_1)$ is negative.

Internal-energy change

$$U_2 - U_1 = mc_v (T_2 - T_1) \quad \text{(Eqn 5.7)}$$

Since $T_1 = T_2$ then $U_2 - U_1 = 0$, that is, there is no change in internal energy.

This is consistent with the molecular model of a perfect gas because heat transfer to the gas increases the linear kinetic energy only and hence the temperature. It follows that if there is no temperature change, there is no internal-energy change.

Heat

Since the process does not take place at constant pressure or constant volume, neither $Q = mc_p (T_2 - T_1)$ nor $Q = mc_v (T_2 - T_1)$ is valid. There is no simple equation for the heat flow, which is best obtained from consideration of the non-flow energy equation.

Non-flow energy equation

$$Q - W = U_2 - U_1$$

Since $U_2 - U_1 = 0$, then $Q = W$, that is, the heat flow equals the work transfer.

Example 5.6

A mass of 0.2 kg of nitrogen at a pressure of 100 kPa (abs.) and a temperature of 300 K is slowly compressed, so that the temperature remains constant, until the pressure reaches 1 MPa (abs.). Calculate:
(a) the initial volume;
(b) the final volume;
(c) the work transfer;
(d) the heat flow.

Solution

Refer to Figure 5.11.

$$m = 0.2 \text{ kg}$$
$$M = 28 \text{ (relative molecular mass of nitrogen)}$$
$$p_1 = 100 \text{ kPa}$$
$$p_2 = 1 \text{ MPa}$$
$$T_1 = T_2 = 300 \text{ K}$$

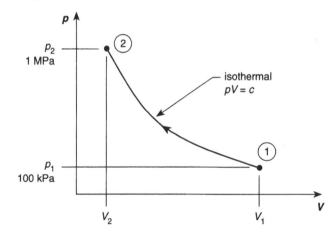

Fig. 5.11

(a) $$p_1 V_1 = m R T_1$$

R for nitrogen may be found from

$$R = \frac{8314}{M} = \frac{8314}{28} = 297 \text{ J/kgK}$$

$$V_1 = \frac{m R T_1}{p_1}$$

$$= \frac{0.2 \times 297 \times 300}{100 \times 10^3}$$

$$= \textbf{0.178 m}^3$$

(b) $p_1V_1 = p_2V_2$

$$\therefore V_2 = \frac{p_1}{p_2} \times V_1$$

$$= \frac{100 \times 10^3 \times 0.178}{1 \times 10^6}$$

$$= \mathbf{0.0178 \ m^3}$$

(c) $W = p_1V_1 \ln \left(\dfrac{V_2}{V_1} \right)$

$$= 100 \times 10^3 \times 0.178 \times \ln \left(\frac{0.0178}{0.178} \right) \ J$$

$$= \mathbf{-41 \ kJ}$$

The negative sign indicates work input to the system.

(d) The work transfer = the heat flow and has the same sign:

$$\therefore Q = \mathbf{-41 \ kJ}$$

The negative sign indicates heat output from the system.

 ## Self-test problem 5.4

1.5 L of air at 350°C and 1250 kPa (abs.) expands isothermally to normal atmospheric pressure. Use the values for air given in Example 5.4. Determine:
(a) the final volume;
(b) the work done;
(c) the heat flow;
(d) the change in internal energy.

5.8   POLYTROPIC PROCESS

In a polytropic process, pressure, volume and temperature all change.

p–V diagram

The equation of the curve is

$$pV^n = c \text{ (a constant)}$$

where n is called the **polytropic index** or the **index of compression** or **expansion**.

The polytropic process looks similar to the isothermal process on the p–V diagram (Fig. 5.12), except that the curves are of different slope depending on the value of n.

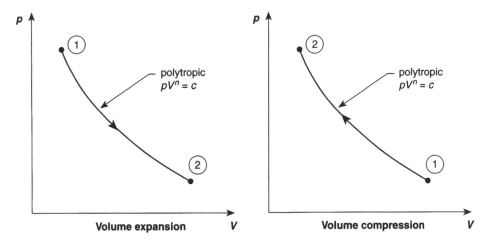

Fig. 5.12 *p–V diagrams for polytropic processes*

Example of the process

An example of a polytropic process is the compression or expansion of a gas under conditions where some heat flow occurs. Most actual compression or expansion processes on gases follow the polytropic process. This is because both compression and expansion do not take place *slowly* enough to allow all the heat to flow (isothermal) or *rapidly* enough to prevent any heat flow at all (adiabatic).

Relationship between *p*, *V* and *T*

Since $pV^n = c$:

$$p_1V_1^n = p_2V_2^n$$

polytropic process (5.13)

Also

$$\frac{T_2}{T_1} = \left(\frac{V_1}{V_2}\right)^{n-1}$$

polytropic process (5.14)

And

$$\frac{p_2}{p_1} = \left(\frac{T_2}{T_1}\right)^{\frac{n}{n-1}}$$

polytropic process (5.15)

Note   Equations 5.14 and 5.15 are derived in Appendix 8.

Work

The work transfer in a polytropic process is given by the following equation (derived in Appendix 9):

$$W = \frac{p_1V_1 - p_2V_2}{n - 1}$$ polytropic work (5.16)

Since $p_1V_1 - p_2V_2 = mR(T_1 - T_2)$, an alternative form is

$$W = \frac{mR}{n - 1}(T_1 - T_2)$$

Heat

As with the isothermal process, the heat flow cannot be obtained directly and is best obtained from the non-flow energy equation.

Example 5.7

After ignition of the fuel mixture at the top of the stroke, an internal-combustion-engine cylinder contains 0.1 L of hot gas at a temperature of 1500°C and a pressure of 7 MPa (absolute). The hot gas expands polytropically ($n = 1.5$) to the bottom of the stroke. The compression ratio is 10:1, $c_p = 1.0$ kJ/kgK and $c_v = 0.72$ kJ/kgK. Determine:
(a) the temperature and pressure at the bottom of the stroke;
(b) the work transfer during the stroke;
(c) the internal-energy change during the stroke;
(d) the heat flow from the cylinder during the stroke.

Solution

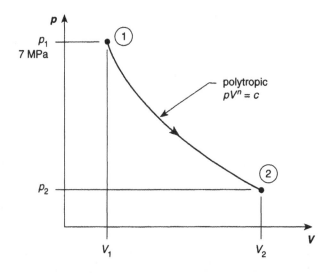

Fig. 5.13

See Figure 5.13, where

$$V_1 = 0.1 \text{ L} = 0.1 \times 10^{-3} \text{ m}^3$$
$$p_1 = 7 \text{ MPa} = 7000 \text{ kPa}$$
$$T_1 = 1773 \text{ K } (1500°C)$$

$$\frac{V_2}{V_1} = 10 \text{ (compression ratio)}$$

$$\therefore V_2 = 1 \text{ L} = 1 \times 10^{-3} \text{ m}^3$$

(a) From Equation 5.14:

$$\frac{T_2}{T_1} = \left(\frac{V_1}{V_2}\right)^{n-1}$$

$$\therefore T_2 = T_1 \left(\frac{V_1}{V_2}\right)^{n-1}$$

$$= 1773 \left(\frac{1}{10}\right)^{1.5-1}$$

$$= \textbf{561 K (288°C)}$$

Also:

$$p_1 V_1{}^n = p_2 V_2{}^n$$

$$\therefore p_2 = p_1 \left(\frac{V_1}{V_2}\right)^n$$

$$= 7000 \times \left(\frac{1}{10}\right)^{1.5}$$

$$= \textbf{221 kPa}$$

(b) From Equation 5.16:

$$W = \frac{p_1 V_1 - p_2 V_2}{n-1}$$

$$= \frac{7000 \times 0.1 \times 10^{-3} - 221 \times 1 \times 10^{-3}}{1.5 - 1} \text{ kJ}$$

$$= \textbf{957 J} \text{ (work out)}$$

(c)
$$U_2 - U_1 = mc_v(T_2 - T_1)$$

Before substituting in this equation, it is necessary to calculate m.
Since $p_1 V_1 = mRT_1$ and $R = c_p - c_v = 0.28 \text{ kJ/kgK}$, then

$$m = \frac{7 \times 10^6 \times 0.1 \times 10^{-3}}{280 \times 1773}$$

$$= 0.001\,41 \text{ kg}$$

$$\therefore U_2 - U_1 = 0.001\,41 \times 0.72 \times (561 - 1773)$$

$$= \textbf{-1.23 kJ} \text{ (decrease in internal energy)}$$

(d) The heat flow may be obtained from the non-flow energy equation:

$$Q - W = U_2 - U_1$$

$$\therefore Q = W + U_2 - U_1$$

$$= +0.957 - 1.23$$

$$= \textbf{-0.273 kJ} \text{ (heat flow out)}$$

Self-test problem 5.5

The compression check of an internal-combustion engine is performed on a cylinder by removing the spark plug and inserting a pressure gauge, then cranking the motor.

Such a test on a new outboard motor shows a pressure at the top of the stroke of 980 kPa (gauge) when the pressure at the bottom of the stroke is 98 kPa (absolute). Assuming compression is polytropic, with $n = 1.2$ and there are no leakage losses, calculate the compression ratio of the engine.

Also, calculate the temperature after compression and the work done per kilogram of air compressed if the temperature of the air at the bottom of the stroke is 45°C.

5.9 *ADIABATIC (ISENTROPIC) PROCESS*

In an adiabatic process, no heat flow occurs. An ideal adiabatic process is an isentropic (constant-entropy) process.

The ideal adiabatic process may be treated as a special case of the polytropic process where the index n is the adiabatic index γ given by:

$$\boxed{\gamma = \frac{c_p}{c_v}}$$
adiabatic index (5.17)

This relationship is derived in Appendix 10.

Note For air, $\gamma = 1005/718 = 1.4$.

p–V diagram

The adiabatic process has the same general shape as the polytropic process and follows the law $pV^\gamma = c$. Refer to Figure 5.12.

Example of the process

In practice, expansion or compression seldom takes place without some heat transfer. However, the adiabatic process is approached when:

- compression or expansion is very rapid (such as the firing of a bullet from a rifle);
- the cylinder and/or piston are made of a non-conducting material such as ceramic, or if the cylinder is insulated as in a steam engine; or if the temperature of the surroundings is approximately equal to that of the gas.

Relationship between *p*, *V* and *T*

Equations 5.13, 5.14 and 5.15 are applicable, but γ is substituted for n.
For instance, Equation 5.13 becomes

$$p_1 V_1^\gamma = p_2 V_2^\gamma$$

Work

Again, Equation 5.16 may be used with $n = \gamma$.

$$W = \frac{p_1 V_1 - p_2 V_2}{\gamma - 1}$$

or

$$W = \frac{mR}{\gamma - 1}(T_1 - T_2)$$

Heat

By definition, $Q = 0$.

Non-flow energy equation

Since $Q = 0$,

$$-W = U_2 - U_1$$

or

$$W = U_1 - U_2$$

Therefore, for the adiabatic process, the work transfer equals the change in internal energy but has the opposite sign (due to the sign convention for work).

Example 5.8

Repeat the calculations for Example 5.7, but assume an ideal adiabatic process.

Solution

Refer to Figure 5.14.

$$c_p = 1.0 \text{ kJ/kgK}$$
$$c_v = 0.72 \text{ kJ/kgK}$$

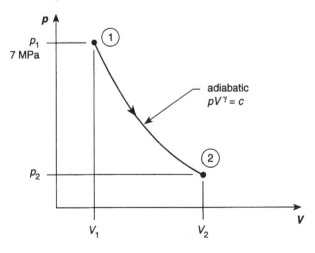

Fig. 5.14

$$V_1 = 0.1 \times 10^{-3} \text{ m}^3$$
$$p_1 = 7000 \text{ kPa}$$
$$T_1 = 1773 \text{ K}$$
$$V_2 = 1 \times 10^{-3} \text{ m}^3$$
$$m = 0.001\,41 \text{ kg}$$

(a)
$$\gamma = \frac{c_p}{c_v}$$

$$= \frac{1.0}{0.72}$$
$$= 1.389$$

$$\frac{T_2}{T_1} = \left(\frac{V_1}{V_2}\right)^{\gamma-1}$$

$$T_2 = 1773 \times \left(\frac{1}{10}\right)^{1.389-1}$$
$$= \textbf{724 K}$$

which is considerably higher than before, since no heat flow takes place. Also,

$$p_1 V_1^{\gamma} = p_2 V_2^{\gamma}$$

$$p_2 = 7000 \times \left(\frac{1}{10}\right)^{1.389}$$
$$= \textbf{286 kPa}$$

which is also higher, as to be expected.

(b)
$$W = \frac{p_1 V_1 - p_2 V_2}{\gamma - 1}$$

$$= \frac{7000 \times 0.1 \times 10^{-3} - 286 \times 1 \times 10^{-3}}{1.389 - 1} \text{ kJ}$$
$$= \textbf{+1.065 kJ}$$

which is also higher, since there is no heat loss.

(c)
$$U_2 - U_1 = mc_v(T_2 - T_1)$$
$$= 0.001\,41 \times 0.72 \times (724 - 1773)$$
$$= \textbf{-1.065 kJ}$$

(d) $Q = 0$ and the non-flow energy equation balances:

$$-W = U_2 - U_1$$

Summary

Calculations involving gases are simplified if the gas is assumed to be perfect or ideal. This is a reasonable assumption in many cases provided that the gas is not under extremes of pressure or temperature or is close to liquefaction point.

With gases the specific heat capacity depends on the nature of the heating process. Two common heating processes are constant-pressure heating and constant-volume heating. The specific heat capacity at constant pressure (c_p) is greater than the specific heat capacity at constant volume (c_v) and the ratio c_p/c_v is called the adiabatic index γ.

Several equations, in particular Equations 5.1 to 5.5, relate to general gas processes and are valid whatever the process on a fixed mass of gas. Other equations are valid only for a particular process. In particular, heat flow and work transfer depend on the type of process. Also if the process is given, property relationships may be simplified to reduce the number of variables. Table 5.1 provides a comparison summary of these processes.

Table 5.1 *Comparison summary of gas processes**

Process	Relationship between p, V, T	Work (W)	Internal-energy change ($U_2 - U_1$)	Heat (Q)
constant pressure	$p = \text{constant}$ $$\frac{V_1}{T_1} = \frac{V_2}{T_2}$$	$p(V_2 - V_1)$	$mc_v(T_2 - T_1)$	$mc_p(T_2 - T_1)$
constant volume	$V = \text{constant}$ $$\frac{p_1}{T_1} = \frac{p_2}{T_2}$$	0	$mc_v(T_2 - T_1)$	$Q = U_2 - U_1$ $Q = mc_v(T_2 - T_1)$
isothermal	$T = c$ $p_1V_1 = p_2V_2$	$p_1V_1 \ln\left(\dfrac{V_2}{V_1}\right)$	0	$Q = W$ $Q = p_1V_1 \ln\left(\dfrac{V_2}{V_1}\right)$
polytropic	$pV^n = \text{constant}$ $$\frac{T_2}{T_1} = \left(\frac{V_1}{V_2}\right)^{n-1}$$ $$\frac{p_2}{p_1} = \left(\frac{T_2}{T_1}\right)^{\frac{n}{n-1}}$$	$\dfrac{p_1V_1 - p_2V_2}{n-1}$	$mc_v(T_2 - T_1)$	$Q = W + U_2 - U_1$
adiabatic	as for polytropic with $n = \gamma$	as for polytropic with $n = \gamma$	$mc_v(T_2 - T_1)$	0

* For all processes, $Q - W = U_2 - U_1$.

Problems

Notes

1 Assume ideal gas behaviour and ideal processes.

2 Gas pressures are absolute unless gauge pressure is stated.

3 Atmospheric pressure is 101.3 kPa.

4 Refer to Appendix 3 for relative molecular masses.

5 For air: c_p = 1005 J/kgK, c_v = 718 J/kgK, R = 287 J/kgK, γ = 1.4

5.1 (a) What are the *three* assumptions necessary in order that a gas be considered perfect?
   (b) Under what conditions do real gases depart appreciably from perfect gas behaviour? Give reasons.
   (c) What two assumptions apply when using the general gas equation, 5.1?
   (d) Explain what is meant by an *isothermal* process and an *adiabatic* process applied to a gas, and give *two* examples of *each*.

5.2 A cylindrical container with hemispherical ends has an internal diameter of 1.2 m and overall internal length of 2.8 m. It is filled with air until a pressure gauge attached to the tank reads 150 kPa. Determine the mass of air in the tank if the air temperature is 15°C.
   8.25 kg

5.3 A cylindrical oxygen-storage bottle has internal diameter 200 mm and length 1600 mm. Determine the gauge pressure reading when the tank contains 0.1 kg of oxygen at 15°C.
   47.6 kPa

5.4 If the oxygen bottle given in Problem 5.3 is recharged with oxygen until the gauge pressure is 800 kPa and the temperature is the same, determine the mass of oxygen added to the cylinder.
   If the oxygen is then used at a constant rate of 0.05 kg/min, calculate the elapsed time until the gauge pressure falls to 100 kPa.
   0.505 kg; 9.4 min

5.5 0.01 kg of air is compressed in a cylinder so that the temperature of the air rises from 20°C to 140°C. If the work done is 2 kJ, calculate the heat flow.
   If the air in the cylinder is now allowed to cool down to 20°C again, while the same cylinder volume is maintained, calculate the additional heat flow.
   −1.14 kJ; −0.862 kJ

5.6 A vertical open-ended cylinder of 80 mm bore and 200 mm length contains air. It is fitted with a close-fitting piston of mass 8 kg. If atmospheric temperature is 15°C, determine the equilibrium position of the face of the piston above the base of the cylinder. Neglect friction.
   To what temperature would the air in the cylinder need to be raised in order to return the piston to the top position?
   173 mm; 59.4°C

5.7 A cylindrical air receiver has an internal diameter of 750 mm and an internal length of 1.5 m. It stores air at a pressure of 850 kPa. Air is used from the tank, and the pressure falls to 700 kPa. Assuming the tank temperature remains at 20°C, calculate the mass of air taken out.
   1.18 kg

5.8 A quantity of gas at a pressure of 120 kPa and temperature 20°C occupies a volume of 3.5 m³. The volume is then reduced to 0.5 m³ and the temperature increased to 100°C. Determine the final pressure.

If the mass of gas is 0.345 kg, what gas is it?

1.07 MPa; hydrogen

5.9 The energy content of a gas is determined using a gas calorimeter and found to be 16.5 MJ/m³. Barometric pressure is 756 mm of mercury and the gas is at a gauge pressure of 20 mm water above atmospheric. Gas temperature is 17.5°C.

Determine the energy content of the gas at normal conditions (20°C and 101.3 kPa). Hint: assume 1 m³ of gas at test conditions and determine what the new volume would be at normal conditions. The total amount of energy does not change.

16.4 MJ/m³

5.10 A cylindrical tank of diameter 5 m and length 7.5 m contains methane (CH_4) at a pressure of 10 kPa (gauge) and a temperature of 20°C. The tank is cooled until the temperature drops to –10°C. Assuming c_p for methane is 2.21 kJ/kgK, determine:

(a) the mass of methane in the tank;

(b) the final pressure shown by the gauge;

(c) the heat flow;

(d) the change in internal energy of the methane.

(a) 107.65 kg   (b) –1.4 kPa   (c) –5.46 MJ   (d) –5.46 MJ

5.11 A pressure gauge on a spherical nitrogen storage tank of internal diameter 2.5 m shows a pressure of 3.5 MPa when the temperature is 25°C. The specific heat capacity at constant pressure for nitrogen is 1040 J/kgK.

(a) Determine the mass of nitrogen in the tank.

If the temperature rises to 34°C, determine:

(b) the new pressure gauge reading;

(c) the heat flow;

(d) the change in internal energy of the nitrogen;

(e) the change in enthalpy of the nitrogen.

(a) 333 kg   (b) 3.61 MPa   (c) 2.23 MJ   (d) 2.23 MJ   (e) 3.12 MJ

5.12 A long vertical cylinder of 100 mm bore, open at the top, contains 0.005 kg of atmospheric air at 15°C. A close-fitting piston of mass 15 kg is placed in the cylinder and allowed to settle to an equilibrium position. Heat energy is then supplied and the piston rises 300 mm. If friction is negligible, determine:

(a) the absolute pressure of the air in the cylinder;

(b) the heat energy supplied;

(c) the change in internal energy of the air inside the cylinder;

(d) the work done by the air in the cylinder.

(a) 120 kPa   (b) 990 J   (c) 708 J   (d) +283 J

5.13 A mass of 0.05 kg of nitrogen at 250°C is heated at a constant pressure of 150 kPa until there is an increase in volume of 100%. Assuming γ for nitrogen is 1.4, determine:

(a) the final volume;

(b) the final temperature;

(c) the specific heat capacity at constant pressure and constant volume;

(d) the heat flow;

(e) the work transfer;

(f) the change in internal energy.

(a) 0.1035 m³   (b) 773°C   (c) 1039 J/kgK, 742 J/kgK   (d) 27.2 kJ   (e) 7.76 kJ
(f) 19.4 kJ

5.14 Five litres of air at atmospheric pressure is slowly compressed through a compression ratio of 10:1 so that the temperature does not change. Determine the work transfer. Also determine the work transfer if the air were compressed so rapidly that negligible heat flow occurred.
   −1.17 kJ; −1.91 kJ

5.15 Two cubic metres of helium at a pressure of 150 kPa and a temperature of 22°C is compressed isothermally through a volume ratio of 4:1. If the adiabatic index for helium is 1.67, determine:
(a) the specific heat capacity at constant pressure and constant volume;
(b) the mass of helium;
(c) the final pressure;
(d) the work done;
(e) the heat flow;
(f) the change in internal energy;
(g) the change in enthalpy.
   (a) 5181 J/kgK, 3102 J/kgK   (b) 0.489 kg   (c) 600 kPa   (d) −416 kJ
   (e) −416 kJ   (f) 0   (g) 0

5.16 Air to 15°C is compressed in a large diesel-engine cylinder from atmospheric pressure to a final pressure of 4 MPa. The index of compression is 1.3. Determine the compression ratio and the temperature at the end of compression.
   16.9:1; 400°C

5.17 Gas is expanded from a pressure of 1.8 MPa and a volume of 0.05 m³ to a pressure of 120 kPa and a volume of 0.38 m³. Assuming expansion follows the law $pV^n = c$, determine the value of n.
   1.335

5.18 Gas at a pressure of 1.4 MPa, temperature 1200°C and volume 0.2 L is expanded in an engine cylinder through a volumetric expansion ratio of 7.4:1. If the index of expansion is 1.3, determine the final volume, pressure and temperature.
   Also, calculate the work done during the expansion.
   1.48 L; 103.8 kPa; 535°C; 421 J

5.19 A mass of 0.05 kg of air at a temperature of 40°C and pressure 100 kPa is compressed to a pressure of 500 kPa, according to the law $pV^{1.3} = c$. Determine:
(a) the final temperature;
(b) the final volume;
(c) the work transfer;
(d) the heat transfer;
(e) the change in internal energy.
   (a) 181°C   (b) 13.02 L   (c) −6.73 kJ   (d) −1.68 kJ   (e) 5.05 kJ

5.20 Repeat Problem 5.19 assuming compression is adiabatic.
   (a) 223°C   (b) 14.2 L   (c) −6.56 kJ   (d) 0   (e) +6.56 kJ

5.21 A reciprocating air compressor has a bore and stroke of 200 mm and a clearance volume of 5% of the swept volume. Maximum pressure of 500 kPa (gauge) is reached and then the delivery valve opens and air is delivered to the receiver. Air enters the cylinder at the

BDC position with a pressure of –10 kPa (gauge). Compression may be considered to be isothermal. Determine for one compression stroke:

(a) the distance moved by the piston from the BDC position when the delivery valve opens;

(b) the work done in compressing the air to maximum pressure;

(c) the heat flow during compression.

    (a) 178 mm   (b) –1.135 kJ   (c) –1.135 kJ

5.22 The piston in an engine cylinder that has a compression ratio of 6:1 compresses 0.2 m³ of air at 20°C and 100 kPa. Heat is then added while the pressure remains the same, until the piston returns to its original position. Compression is polytropic, with $n = 1.3$.

Sketch these two processes on a *p–V* diagram, and determine:

(a) the mass of air;

(b) the pressure at the end of the compression;

(c) the final temperature;

(d) the nett work transfer for the two processes.

    (a) 0.238 kg   (b) 1.027 MPa   (c) 2736°C   (d) 123.7 kJ

5.23 Air at a pressure of 1 MPa and a temperature of 130°C is expanded adiabatically to a pressure of 105 kPa and then compressed isothermally to its original volume. Draw these processes on a *p–V* diagram and determine:

(a) the initial specific volume;

(b) the final temperature;

(c) the final pressure;

(d) the change in internal energy per kilogram;

(e) the nett work transfer per kilogram;

(f) the nett heat flow per kilogram.

    (a) 115.7 L/kg   (b) –61.3°C   (c) 525 kPa   (d) –137 kJ   (e) 39.6 kJ   (f) –97.8 kJ

5.24 A quantity of carbon dioxide occupies a volume of 0.14 m³ at 975 kPa and 370°C. It is heated at constant volume to a pressure of 4.2 MPa, after which it is expanded in an insulated cylinder to a pressure of 270 kPa. Sketch these processes on a *p–V* diagram. Assuming $c_p = 1.075$ kJ/kgK for carbon dioxide, determine:

(a) the temperature at the start of the expansion;

(b) the volume after expansion;

(c) the temperature after expansion;

(d) the nett work transfer;

(e) the nett heat flow;

(f) the nett change in internal energy.

    (a) 2497°C   (b) 1.344 m³   (c) 1437°C   (d) 1.06 MJ   (e) 2.12 MJ   (f) 1.06 MJ

5.25 A mass of 1 kg of air at 100 kPa and 20°C is compressed isothermally to a pressure of 400 kPa. 100 kJ of heat are then added at constant pressure and the gas is then expanded adiabatically back to its original pressure. Sketch these processes on a *p–V* diagram (not to scale) and determine:

(a) the work transfer during compression;

(b) the temperature after the heat is added;

(c) the work transfer during adiabatic expansion;

(d) the final volume.

Now resketch the *p–V* diagram to approximate scale.

    (a) –116.6 kJ   (b) 119.5°C   (c) 92.1 kJ   (d) 0.758 m³

CHAPTER 6

Heat engines

≡

Objectives

On completion of this chapter you should be able to:

- define a heat engine and describe the essentials of a heat engine;
- use the conservation of energy principle to perform an energy balance for a heat engine (considered as a black box);
- recognise that heat engines have an efficiency limit and calculate this limiting efficiency (Carnot efficiency);
- describe the various heat sources used for heat engines;
- describe the various working substances used with heat engines;
- describe the various mechanical arrangements used with heat engines and, in particular, the difference between positive and non-positive displacement and linear and rotary configurations;
- describe the various working cycles used with heat engines and, in particular, the difference between open and closed cycles, internal and external combustion, compression ignition and spark ignition, two-stroke and four-stroke operation;
- draw a theoretical p–V diagram and use it to explain the operation of the following heat engine cycles: Carnot, Otto, Diesel, dual, two-stroke (spark ignition and compression ignition) and Joule cycles—also describe the likely points of difference between the theoretical cycle and the actual cycle.

Introduction

In this chapter, the heat engine is defined and the essentials of a heat engine are described. A systems analysis applicable to *all* heat engines is given (black-box analysis). The need for a heat engine to reject heat and the effect of this on the limiting efficiency of heat engines are treated. A classification of heat engines according to heat source, working substance, mechanical arrangement and working cycle is presented. The important heat-engine cycles are given, with appropriate p–V diagrams and the variation between each theoretical and actual cycle is discussed. The performance and testing of heat engines is given in Chapter 7.

6.1 *DEFINITION OF A HEAT ENGINE*

A heat engine is a device that continuously converts heat to work (or power). The word 'continuous' is of critical importance in this definition and needs further elaboration; it means that the engine will continue to operate as long as the heat energy input is maintained. The ability of a device to convert heat to work does not necessarily make the device a heat engine, as will be seen from consideration of Figure 6.1. Here, gas under pressure is heated and the piston moves toward the right; the opposing force moves through a distance and work is done. The energy that has been converted to work has come from the source of heat, so heat has been converted to work.

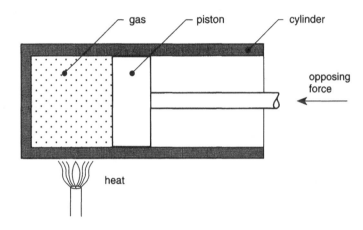

Fig. 6.1 *A device that converts heat into work but is not a heat engine*

However, this device is not a heat engine because it cannot operate *continuously*. One reason for this is the finite length of the cylinder. But this appears only to be a practical limitation—what if the cylinder were infinitely long? Even then the piston could travel only a finite distance before stopping. This is because there is a finite mass of gas enclosed and the gas cannot be heated to a greater temperature than the temperature of the heat source. So at best, the pressure will remain constant for a while as the temperature of the gas increases and the piston moves along the cylinder. Once the temperature of the gas reaches source temperature, further movement of the piston can only be achieved by reducing the opposing force. Hence the gas pressure will fall until eventually no nett force is exerted on the piston and no further work is done. The p–V diagram for this sequence of processes is shown in Figure 6.2 on page 120.

It is therefore apparent that this device can convert heat to work, but that it cannot do so continuously. In order to operate continuously, it is necessary for the piston to return to its initial position. Furthermore, the state of the gas must be the same at this point. For example, if the pressure were lower, the p–V diagram for the next sequence of operations would look the same as Figure 6.2 except that the line would lie below the original line. The next time it would be lower again, and so on, until the engine would once again stop. Therefore, for continuous operation, the processes must constitute a sequence that returns the system to its original condition, that is, there must be a **cycle**.

This leads to an alternative definition: *a heat engine is a device that operates in a cycle, and in so doing, converts heat to work.*

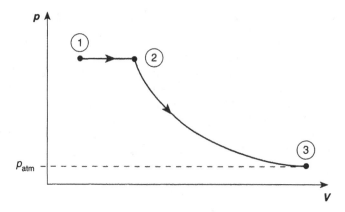

Fig. 6.2 *p–V diagram for the device shown in Figure 6.1*

Example 6.1

Discuss the following devices with regard to the three criteria: heat input, work output and continuous or cyclic operation and hence whether or not they are heat engines:

(a) a shell-firing cannon;
(b) an aircraft jet engine;
(c) a steam turbine;
(d) an electric motor;
(e) a solar-powered motor-car drive mechanism;
(f) a hot-water heater;
(g) a bimetal rod, which rotates when exposed to sunlight.

Solution

(a) A cannon that fires a shell is not a heat engine because there is no working cycle.
(b) An aircraft jet engine is a heat engine. The useful work output occurs when the engine itself moves forward because of the thrust developed.
(c) A steam turbine (or for that matter *any* turbine) is not itself a heat engine because no heat source is used. However, if the steam is generated in a boiler where heat energy is used, then the total system including the boiler is a heat engine.
(d) Electric motors are not heat engines because the electrical energy input is not heat. In fact, heat due to internal resistance is a loss of energy in an electric motor.
(e) A solar-powered motor-car drive mechanism is a heat engine because there is a mechanical power output and it will continue to operate as long as the heat source (solar energy) remains.
(f) A hot-water heater (whether solar or electric) is not a heat engine because there is no work output.
(g) A bimetal rod, which rotates when exposed to sunlight, satisfies the three criteria and is a heat engine. In this case, the work output occurs as a result of the expansion of a solid rather than a fluid.

Need for a heat engine to reject heat

A fundamental requirement of a heat engine is that, at some stage in the cycle, heat must be rejected, that is, transferred from the engine to the surroundings.

For example, it has been shown that the device illustrated in Figure 6.1 is not a heat engine because it does not operate in a cycle. In order to be a heat engine, it is necessary to return the gas to the same condition that it had at ①. How could this be done? An obvious solution is to retrace the path back again from ③ → ② → ① by compressing the gas from ③ → ② and then removing the heat source and cooling the gas at constant pressure from ② → ①. The device would now operate in a cycle but it would still not be a heat engine because no nett work would be done. The same amount of work would be needed to reverse conditions as was originally obtained (in fact more because of friction).

Therefore, in order to have a useful work output, the return path must lie below the original path on the *p–V* diagram. This gives an enclosed area equal to the nett work output per cycle. One such path is shown in Figure 6.3.

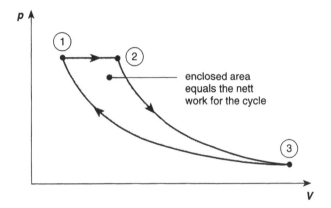

Fig. 6.3 *Nett work output occurs from the cycle when there is a positive closed loop on the p–V diagram*

The only way that this condition can be fulfilled is if, during some part of the return path, heat is rejected, that is, heat flows away from the gas to the surroundings. It is an instructive exercise to use the various gas processes given in Chapter 5 to try to devise a cycle that has a nett work output without rejecting heat. It will be found to be impossible. Hence one of the essential requirements of a heat-engine cycle is that at least one of the processes that constitute the cycle must be one in which heat is rejected.

6.2   *ESSENTIALS OF A HEAT ENGINE*

The five essentials of a heat engine are: a heat source, a working substance, mechanical work (or power) output, a working cycle and a heat sink.

A heat source   A heat engine converts energy (in the form of heat) to mechanical work and therefore must have an energy source in the form of heat.

A working substance   In order to convert heat to work, a substance is required that expands when heated and thereby is able to move a force through a distance and provide work. Gases (or vapours) do this most effectively and hence the vast majority of heat engines use these as the working substance. In some cases, a phase change may occur (for example water–steam–water in steam engines). A solid working substance is also possible (for example the bimetal rod, which rotates when exposed to sunlight).

Because the working substance is usually a fluid, the working substance is also known as the working fluid.

Mechanical work (or power) output   This is usually in the form of rotating-shaft power but linear power output is also used.

A working cycle   A heat engine cannot operate continuously unless the working substance goes through a series of processes linked together in such a way that the system returns to its original condition and there is a nett work output. That is to say, the processes must form a loop on a *p–V* diagram that has a positive area. The series of processes is known as the working cycle.

A heat sink   At least one process in the working cycle must be one in which heat is rejected to the surroundings. Because heat will flow only across a temperature difference, this means that the engine requires a heat sink that is able to absorb the rejected heat and is at a lower temperature than the source. On earth, the heat sink for heat engines is the atmosphere, from which the energy radiates out into space. Usually atmospheric air provides the immediate heat sink but in other cases lakes or the ocean may be used for better initial heat transfer.

6.3   *EFFICIENCY OF A HEAT ENGINE*

Heat engines may be analysed by black-box-systems analysis, as shown in Figure 6.4.

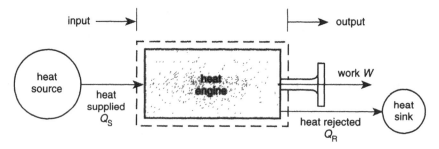

Fig. 6.4   *Systems analysis of a heat engine regarded as a black box*

Under steady-state conditions over a certain time period:

$$Q_S = W + Q_R$$

or

$$\boxed{W = Q_S - Q_R}\quad \text{work output of a heat engine (6.1)}$$

Using a time period of 1 s:

$$\boxed{P = \dot{Q}_S - \dot{Q}_R}\quad \text{power output of a heat engine (6.2)}$$

The efficiency is:

$$\boxed{\eta = \frac{W}{Q_S} = \frac{P}{\dot{Q}_S}}\quad \text{efficiency of a heat engine (6.3)}$$

that is

$$\eta = \frac{Q_S - Q_R}{Q_S}$$

or

$$\boxed{\eta = 1 - \frac{Q_R}{Q_S} = 1 - \frac{\dot{Q}_R}{\dot{Q}_S}}$$ efficiency of a heat engine (6.4)

where Q_S = quantity of heat supplied (J)
        Q_R = quantity of heat rejected (J)
        W = work output (J)
        $\dot{Q}_S$ = heat-supply rate (W)
        $\dot{Q}_R$ = heat-rejection rate (W)
        P = power output (W)

Notes

1 This efficiency may also be termed **overall efficiency** or **thermal efficiency** because it is the efficiency with which heat energy (thermal energy) is converted to mechanical energy. The output power is usually measured by some form of engine brake, so this efficiency is also known as the **brake thermal efficiency**.

2 In these equations the signs are according to the convention and have already been included in the equations. They should not be included again when substituting in the equations; for example, if Q_R is taken as negative when substituting in Equation 6.1, the result is $W = Q_S - (-Q_R) = Q_S + Q_R$ which is incorrect.

Self-test problem 6.1

A heat engine uses fuel of energy content 43.1 MJ/kg and produces 17.4 kW of useful power. The heat-rejection rate (through the exhaust and cooling systems) is 44.8 kW. Determine:
(a) the efficiency of the engine;
(b) the usage rate of fuel in kg/h.

6.4 *MAXIMUM EFFICIENCY OF A HEAT ENGINE*

From Equation 6.4, it is evident that the efficiency must always be less than 1 (100%), because Q_R must have some value. The questions are:
(a) Is there some limit to the efficiency of a heat engine?
(b) How close to 100% can the efficiency be?
(c) What determines the maximum efficiency? For example, can it continually be brought closer to 100% by better design of components, or by reducing friction, or by using more sophisticated materials such as ceramics, or by improving the working substance?

These questions were first analysed by Sadi Carnot in 1824. He came to the following conclusions:

• There is a limit to the efficiency of a heat engine. No matter how well the engine is constructed (even if friction and heat losses could be eliminated entirely), the engine has a limit to the attainable efficiency because of the need to reject heat during some part of the cycle.

- The limit to the efficiency depends on the difference between the temperature of the working substance when heat is accepted and the temperature when heat is rejected. The greater this temperature difference, the greater the limiting efficiency, and the lower the temperature difference, the lower the limiting efficiency.

The limiting efficiency (or maximum possible efficiency) of a heat engine is known as the **Carnot efficiency** and is given by the following equation (derived in Appendix 11).

$$\eta_C = \frac{T_h - T_c}{T_h}$$

or

$$\boxed{\eta_C = 1 - \frac{T_c}{T_h}}$$

Carnot efficiency (6.5)

where T_h = temperature in kelvins of the working substance when heat flows *in* from the source (*hot* temperature)

T_c = temperature in kelvins of the working substance when heat flows *out* to the sink (*cold* temperature)

Notes

1 The maximum and minimum temperatures of the working substance are, in turn, governed by the temperature of the source and the temperature of the sink.

2 The limit to the efficiency of a heat engine does not apply to a single process of converting heat to work or work to heat. For example, it was shown in Chapter 5 that for an isothermal process, heat may be converted to work with 100% efficiency (if there is no friction). The conversion of work to heat is achieved with 100% efficiency by a stirrer or paddle. The limit to the efficiency of a heat engine occurs when the various processes are joined together to form a working cycle.

3 The Carnot efficiency is the limiting efficiency. This means that if the efficiency of an engine is plotted on a graph with efficiency on the vertical axis, the 100% line is not applicable and a horizontal line should be drawn through the Carnot efficiency. The heat engine will then operate somewhere in the range of efficiency $0–\eta_C$.

4 A working heat engine will always have an efficiency less than η_C. This is due to friction (both mechanical and fluid), turbulance, vibration (sound), leakage of fluid (mass loss) and heat loss. There must be a temperature difference between the source and the working substance for heat to flow in; therefore the working substance cannot achieve maximum source temperature. Similarly, the working substance cannot achieve minimum sink temperature. Also, unless the heat transfer occurs isothermally, not all the heat is transferred *to* the working substance at maximum temperature or *from* the working substance at minimum temperature.

All these factors combine to lower the efficiency of a working heat engine to considerably below the limiting efficiency. A heat engine with an actual overall efficiency of 50% may be considered to be a highly efficient engine, and a value of 20–30% is far more common.

Example 6.2

Determine the maximum possible efficiency of the following heat engines:

(a) an internal-combustion engine using the atmosphere at 20°C as the heat sink, the temperature after combustion being 1450°C;

(b) a steam-cycle power station using a supercritical one-through boiler with a steam temperature of 750°C and a condensing temperature of 20°C;

(c) a steam-cycle power station using a conventional drum-and-tube boiler with a steam temperature of 570°C and a condensing temperature of 20°C;

(d) a power station designed to work on the temperature difference in sea water, the cold temperature being 10°C and the hot temperature 22°C.

Solution

(a) $T_c = 293$ K, $T_h = 1723$ K

$$\eta_C = 1 - \frac{293}{1723}$$
$$= \mathbf{0.83}$$

(b) $T_c = 293$ K, $T_h = 1023$ K

$$\eta_C = 1 - \frac{293}{1023}$$
$$= \mathbf{0.71}$$

(c) $T_c = 293$ K, $T_h = 843$ K

$$\eta_C = 1 - \frac{293}{843}$$
$$= \mathbf{0.65}$$

(d) $T_c = 283$ K, $T_h = 295$ K

$$\eta_C = 1 - \frac{293}{295}$$
$$= \mathbf{0.04}$$

 ## Self-test problem 6.2

Draw a graph of Carnot efficiency against upper-cycle temperature for upper-cycle temperatures in the range 15–2000°C. The lower-cycle temperature may be taken to be the atmosphere at 15°C. Use upper-cycle temperatures of 15, 100, 200, 300, 500, 1000, 1500 and 2000°C.

6.5 TYPES OF HEAT ENGINE

The various types of heat engines may be classified by the working substance, heat source, mechanical arrangement, and working cycle as shown in Figure 6.5 on page 126.

The heat source The majority of heat engines use combustion of a fuel (chemical energy) in order to provide the input heat energy. Most commonly, liquid fuels are used because of their convenience. Alcohol and petroleum derivatives such as petrol, diesel fuel, fuel oil and kerosene are suitable liquid fuels. Gaseous fuels such as natural gas or hydrogen may also be used. Solid fuels (in particular coal) are used with steam engines or steam turbines, particularly for generating electrical power.

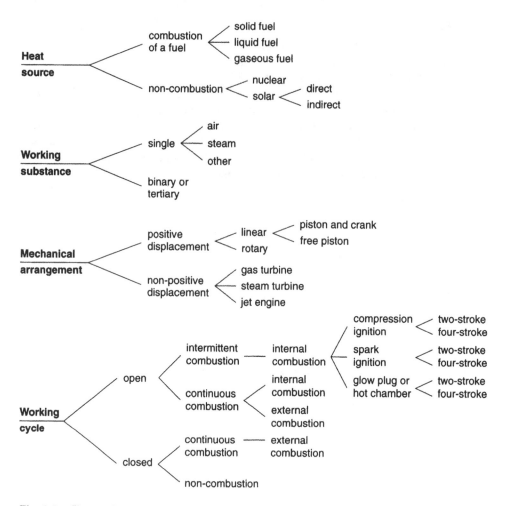

Fig. 6.5   *Types of heat engines*

As discussed in Chapter 1, fossil fuels are in fixed supply and will eventually be used up. The two main replacement heat-energy sources are nuclear and solar energy. Both are feasible but both have their difficulties.

The working substance   Most heat engines use a single working substance. Since air and water are readily available (and cheap) most heat engines use air or water–steam as the working substance. In some cases, it it advantageous to have two (binary) or more (tertiary) working substances. With two or more working substances, heat exchanges are used to transfer the heat from one substance to another.

It is also possible to use a solid working substance but this has novelty value only.

The mechanical arrangement   The two main types of mechanical arrangements are positive and non-positive displacement (rotodynamic). In positive-displacement engines, the working fluid is physically displaced (for example by a piston moving in a cylinder). In non-positive-displacement (or rotodynamic) engines, the displacement of the working fluid is caused by pressure differences (for example by a fan or turbine).

Positive-displacement engines In these engines the motion may be linear or rotary. The most common linear-motion mechanism is by means of a reciprocating piston in a cylinder driven by a crankshaft and connecting rod. Single-cylinder or multicylinder configurations are used in many variations such as in-line, vee, opposed (single or double crank) or radial. Typical configurations are illustrated in Figure 6.6.

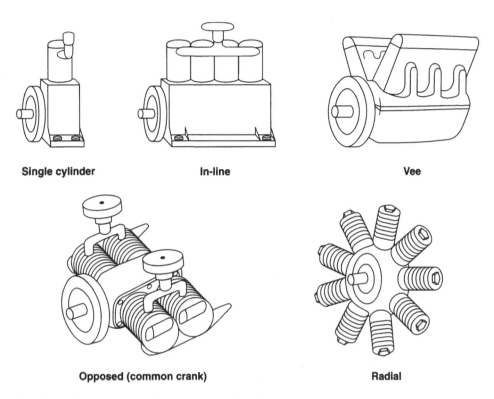

Single cylinder	In-line	Vee

Opposed (common crank)   Radial

Fig. 6.6 *Some reciprocating engine configurations*

Because the torque on the crankshaft produced by a single reciprocating piston varies throughout the working cycle, a flywheel is usually necessary to smooth out the torque variations and reduce the vibration. The size of the flywheel (and the weight of the engine for a given power output) can be reduced by having a number of cylinders fitted with pistons that are out of phase with each other. A greater number of cylinders also means that each cylinder has a smaller combustion chamber. This allows the fuel to burn more rapidly so that the engine is able to run at higher speeds, which also increases the power output available from a given engine capacity.

As discussed in Chapter 2, the mechanism may be single acting or double acting. The double-acting configuration is usually preferred for steam engines and large slow-revving diesel engines (such as used in large ships). The single-acting configuration is used in most other applications.

The reciprocating piston mechanism provides good sealing but has obvious disadvantages. For example at 6000 rpm, the piston is accelerating from rest and decelerating back to rest again 200 times per second! Rotary, positive-displacement engines have been the subject of intense research and development and several designs have been developed (for example the Wankl) but have not replaced the conventional reciprocating mechanism.

Positive-displacement engines may also be of the free-piston type, where there is no crankshaft and connecting rod. The work output of the piston is used to directly pump or compress another fluid, but such engines are rare.

Non-positive-displacement heat engines   These are of three main types, namely steam turbines, gas turbines and jet engines.

In a steam turbine, the steam is generated in a boiler before passing through the turbine. Gas turbines and jet engines use a rotary compressor and turbine, between which is a combustion chamber. The main difference between gas turbines and jet engines is that in a gas turbine, the output power is in the form of shaft power whereas in a jet engine all or most of the output power is in the form of thrust from the hot exhaust gas.

Jet engines are used almost exclusively on aircraft. Three main types of jet engines are used depending on the speed and altitude of the aircraft. In **pure jet** engines, all the output power is in the form of thrust from the exhaust gas and these engines are used for high-speed, high-altitude applications. In **prop-jet** engines, the majority of the output power is in the form of shaft power, which is used to drive a large, external propeller. These engines are used for low-speed, low-altitude applications. **By-pass** or **fan-jet** engines are an intermediate type, where an internal fan is coupled to the shaft. This fan acts in a similar way to a propellor and provides additional thrust.

Turbines whether steam or gas may derive their thrust from either impulse force, reaction force or a combination of both. In **impulse turbines**, force (and power) is developed from a *change in direction* of the working fluid as it passes across the moving turbine blades. In **reaction turbines**, the force (and power) is developed from the *acceleration* of the working fluid in the moving blades. These different methods are illustrated in Figure 6.7.

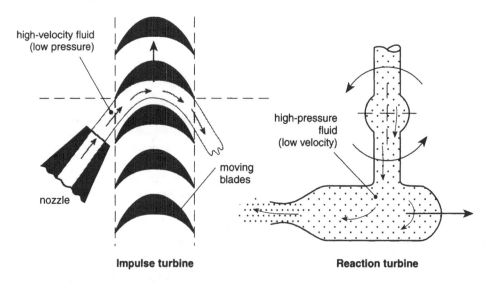

Fig. 6.7   *Impulse and reaction turbines*

Many turbines use both principles of impulse and reaction, and so may be termed **impulse–reaction turbines**. Part of the power output is obtained from the change in direction of the fluid in the moving blades and part from the acceleration of the fluid in the moving blades (see Fig. 6.8). The acceleration is achieved by reducing the cross-sectional area of the fluid path in the blades. However, the fluid will only flow through the constriction if there is a pressure difference across the blades. This leads to the following definition: *Impulse turbines have no pressure drop (apart from that due to fluid friction) across the moving blades, whereas impulse–reaction and reaction turbines have a pressure drop across the moving blades.*

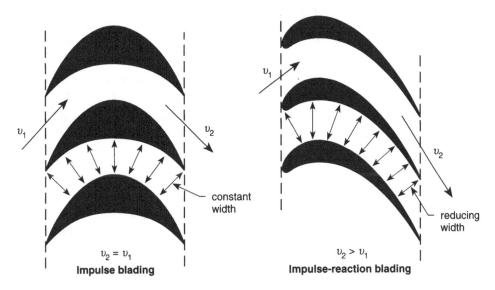

Fig. 6.8　*Blading for impulse and impulse-reaction turbines*

The working cycle

The two main types of working cycle are open and closed cycles.

Open cycles　The working substance flows through the engine. The working substance is usually air although a steam engine may also operate on an open cycle.

Closed cycles　The working substance remains within the engine, circulating around it without being replenished (except for leakage losses). Closed-cycle engines may use any working substance.

There are other variations to the open or closed cycles that distinguish the type of engine. Some of these distinctions are as listed below.

Continuous combustion　Heat is supplied from a fuel that burns continuously.

Intermittent combustion　The fuel does not burn continuously but is ignited periodically at the appropriate point in the cycle.

Internal combustion　The air used to sustain combustion is also used as the working substance, that is, combustion takes place internally (within the combustion chamber of the engine).

External combustion　The air needed for combustion is not used as the working substance. The engine may use virtually any working substance. External combustion requires that the heat of combustion is transferred to the working substance through the cylinder wall or by means of a heat exchanger.

Spark ignition   Energy in the form of an electrical spark is used to initiate combustion.

Compression ignition   The air used as the working substance is compressed to a temperature high enough to allow the fuel to burn when injected into the combustion chamber. This means that compression-ignition engines do not need auxiliary ignition equipment such as spark plugs.

Glow-plug or hot-chamber ignition   Combustion is initiated by the heat provided by a glow plug or hot chamber. This is a method intermediate between compression ignition and spark ignition and is not very widely used nowadays (except in model aircraft engines).

Two-stroke   This is a term used with reciprocating engines to indicate that the cycle is completed every two strokes of a given piston, that is, for one revolution of the crank.

Four-stroke   This is a term used with reciprocating engines to indicate that the cycle is completed every four strokes of a given piston, that is, for two revolutions of the crank.

 ## *Self-test problem 6.3*

Outline the main advantages and disadvantages of continuous combustion compared with intermittent combustion, and external combustion compared with internal combustion in heat engines.

6.6   *CARNOT CYCLE*

When considering heat-engine cycles the question that may be asked is: Is there a heat-engine cycle that has an efficiency equal to the maximum possible efficiency and, if so, what series of processes would constitute this cycle?

In order to obtain maximum possible efficiency, all the heat transferred in must be at the maximum cycle temperature T_h, and all the heat transferred out must be at the minimum cycle temperature T_c. The process by which transfer of heat across a system boundary occurs at constant temperature is the isothermal process. If, for any other processes, no heat transfer across the system boundary occurs, then these processes will be adiabatic. The combination of ideal adiabatic and isothermal processes to form a working cycle is known (naturally enough) as the **Carnot cycle**, and the p–V diagram for this cycle is shown in Figure 6.9.

The following processes constitute the Carnot cycle:

① → ② isothermal compression; heat rejected to sink at constant temperature T_c; the pressure increases

② → ③ adiabatic compression; the pressure and temperature both increase without any heat flow taking place

③ → ④ isothermal expansion; heat supplied from source at constant temperature T_h; the pressure decreases

④ → ① adiabatic expansion; the pressure and temperature both decrease until they have their same initial values; no heat flow occurs

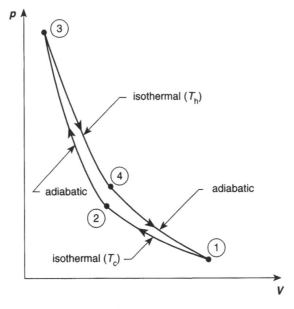

Fig. 6.9 *p–V diagram for a Carnot cycle*

A theoretical engine operating on this cycle would be as shown in Figure 6.10 where a source, sink and insulator are alternately brought into contact with the working substance.

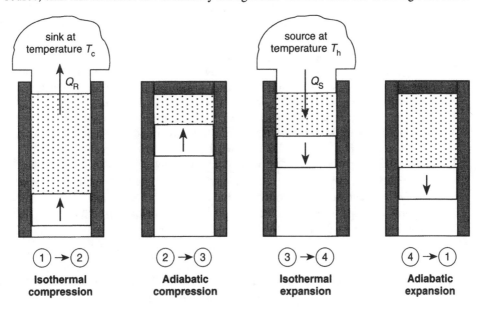

Fig. 6.10 *Reciprocating heat engine operating on the Carnot cycle*

The efficiency of the engine is given by the Carnot efficiency (Equation 6.5) derived in Appendix 11.

$$\eta_C = 1 - \frac{T_c}{T_h}$$

In practice, the Carnot cycle cannot be utilised directly for a working heat engine because the slopes of the isothermal and adiabatic lines on the *p–V* diagram are not sufficiently different to enclose an area great enough to provide a useful work output with the friction and other losses present in an actual engine.

6.7 *STIRLING CYCLE*

The Stirling cycle is an external-heat, closed two-stroke cycle, which uses a gas as the working substance. It was invented by Robert Stirling in about 1816 and has historical importance as one of the first practical heat engines to use air rather than steam as the working substance. The basic principle of the engine is that one part of a cylinder (or a separate cylinder) is kept hot by the application of the heat source (usually by external combustion of a fuel), and another part of the cylinder (or separate cylinder) is kept cool by the use of cooling fins or water cooling, and is the heat sink.

The working substance (gas) is compressed in the cold space and heat is rejected. The gas is then transferred to the hot space, where it expands and does work. The gas is then returned back to the cold space and the cycle repeats.

The transfer of the gas from the cold space to the hot space and back again is usually achieved by having a displacer or displacer piston in addition to the power piston, the motion of which is synchronised to the movement of the power piston. The distinction between a displacer and a piston is that a displacer is a very loose fit in the cylinder so that the gas can readily flow around it whereas a piston provides a gas-tight seal.

The Stirling cycle is a regenerative cycle, which uses a regenerator to alternatively absorb and give back heat as the gas flows over it. Because the regenerator is internal, external heat loss is minimised and better efficiency is obtained. One form of Stirling engine uses a porous matrix regenerator but in other cases the displacer itself is designed to serve as a regenerator. In such cases, the displacer is long and has a thin wall. The length of the displacer ensures that despite the reciprocating movement of it, one end is always in the hot section whereas the other end is always in the cold section. The thin-wall design minimises the conduction of heat along the length of the displacer.

The Stirling cycle may be adapted to a great variety of configurations and designs, many of which have been constructed and tested. Two basic configurations are shown in Figure 6.11 (opposite), as well as the theoretical and actual *p–V* diagrams. Following the points marked on the theoretical cycle and with the displacer type of engine (a):

①   Displacer at TDC, power piston at BDC.

① – ②   The piston moves up and compresses the gas isothermally in the cold space. Heat flows out to the heat sink.

② – ③   The displacer moves down and transfers the compressed gas to the hot space. The regenerative function of the displacer means that the gas is also heated regeneratively during this transfer operation. The volume remains constant but the pressure increases.

③ – ④   The power piston moves down and does work. Heat transfers into the gas from the heat source so that the temperature remains constant (isothermal).

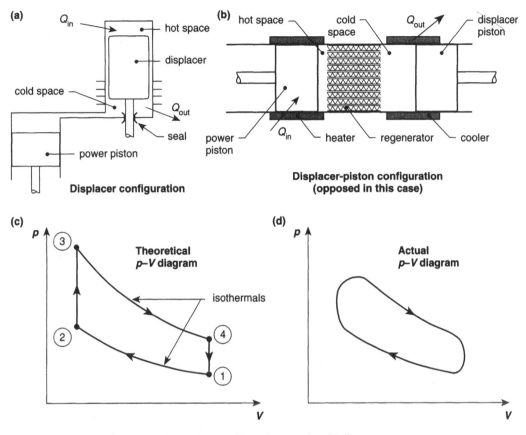

Fig. 6.11   *Stirling-cycle engine configurations and p–V diagrams*

④ – ①   The displacer moves up and transfers the hot gas through the regenerator to the cold space. The temperature of the gas falls as heat transfers to the regenerator so that the pressure also falls but the volume remains constant.

①   Displacer at TDC, power piston at BDC. The cycle may now repeat again.

In the theoretical cycle described above, the piston and displacer are at rest during phases of the cycle. In practice this is difficult to do and, in an operating engine, they are linked together usually by a crankshaft-connecting rod arrangement and move continuously. This means that the *p–V* diagram for an operating engine does not exhibit the sharp corners of the theoretical cycle so that the diagram has an almost elliptical shape.

Efficiency of the Stirling cycle

In theory, heat transfers from heat source to working substance to heat sink occur isothermally. No other external heat transfers occur. Therefore the Stirling cycle has the same theoretical efficiency as the Carnot cycle as given by Equation 6.5:

$$\eta = 1 - \frac{T_c}{T_h}$$

Advantages and disadvantages of the Stirling cycle

Advantages

Stirling-cycle engines offer some remarkable advantages:

- Virtually any heat source (including solar) may be used. If combustion of a fuel is used as the heat source, any combustible material may be utilised.
- Because combustion is external, and continuous, it can be carefully controlled. Excess air may be used and pollutants can be kept to a minimum.
- The engine is very quiet, with little vibration.
- The engine may be run in reverse (with power input rather than output) and used as a heat pump *or* refrigerator.
- The engine needs very little sophisticated equipment, which no doubt explains why it was one of the first practical heat engines to be constructed. An operating engine can be constructed without the need for valves, camshafts, spark plugs, electrical equipment, fuel injectors and such like. This gives the engine the potential to be very reliable with little maintenance needed.

Disadvantages

Despite the powerful advantages, and intense research and development, Stirling engines have not achieved significant popularity. This is because there are also some major disadvantages associated with them:

- Efficiency and power output per unit mass (specific power output) are lower than is the case with well-designed internal-combustion engines. This is mainly because the working substance cannot be raised to as high a temperature because the heat has to transfer through the walls of the cylinder. Typically, the maximum temperature is of the order of about 700°C whereas in an internal-combustion engine, temperatures of two to three times this figure are achieved. Also, because of the time it takes for heat to transfer through metals, and because of the necessity of transferring the gas forward and backward through the regenerator, Stirling-cycle engines cannot achieve the high speeds and therefore high power outputs of faster revving engines. Efficiency and power output can be improved by using compressed hydrogen or helium as the working substance but this introduces formidable difficulties with regard to sealing and make-up of gas losses.
- Despite the basic simplicity of the cycle, high-performance Stirling engines are more complex and costly than internal-combustion engines (per kW of output power). This is because more expensive materials are needed in critical components such as the high-temperature heat exchanger and regenerator. If a pressurised gas (other than air is used) sealing becomes a major difficulty (and cost).
- Cooling is difficult and requires a large heat exchanger. In an internal-combustion engine, much of the rejected heat goes out in the exhaust gas (as well as the cooling system) but in a Stirling engine all the rejected heat has to be transferred through a heat exchanger. Furthermore, in order to obtain a high efficiency, the temperature at which the heat is rejected needs to be low and this is difficult to achieve with conventional cooling systems.
- A warm-up time is necessary from cold-start conditions to heat the hot space to working temperature.

Example 6.3

In a Stirling-cycle engine, the working substance achieves a maximum temperature of 600°C and a minimum temperature of 100°C. The heat supply rate is 50 kW and the actual efficiency is 25%. Determine:

(a) the ideal efficiency;
(b) the ideal power output;
(c) the actual power output;
(d) the actual heat rejection rate.

Solution

(a)
$$\eta_{ideal} = 1 - \frac{T_c}{T_h}$$
$$= 1 - \frac{373}{873}$$
$$= \mathbf{0.573}$$

(b)
$$P_{ideal} = \eta_{ideal} \times \dot{Q}$$
$$= 0.573 \times 50$$
$$= \mathbf{28.6\ kW}$$

(c)
$$P_{actual} = \eta_{actual} \times \dot{Q}$$
$$= 0.25 \times 50$$
$$= \mathbf{12.5\ kW}$$

(d)
$$\dot{Q}_R = \dot{Q}_S - P$$
$$= 50 - 12.5$$
$$= \mathbf{37.5\ kW}$$

Self-test problem 6.4

A solar-powered Stirling engine uses a focusing mirror of diameter 1.2 m. This enables the working fluid to achieve a maximum temperature of 360°C on a day when solar radiation received is 600 W/m² and air temperature is 20°C. Determine the following theoretical values:

(a) the efficiency;
(b) the rate of heat supply;
(c) the power output;
(d) the rate of heat rejection.

6.8 OTTO CYCLE

The Otto cycle is a positive-displacement, four-stroke, internal-combustion, spark-ignition, open cycle, invented by Nickolous Otto in the 1860s.

The sequence of the process in this cycle in a reciprocating engine and the theoretical and actual p–V diagrams are shown in Figure 6.12 (page 136) and should be self-explanatory. Not shown is the camshaft, which controls the valve movements and the carburettor or fuel injectors, which provide the correct air–fuel mixture. Petrol is by far the most predominant fuel for these engines so they are also known as petrol engines. However, gaseous fuels may also be used successfully.

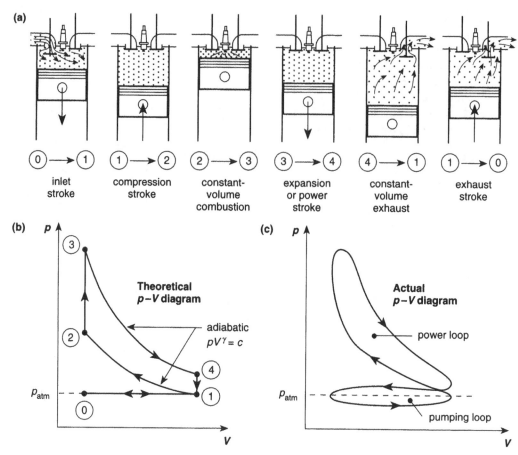

Fig. 6.12 *Otto-cycle reciprocating-engine sequence and p–V diagrams*

The main difference between the actual and theoretical $p–V$ diagrams are as follows:

- The actual diagram has rounded rather than sharp corners. This is because valve movement and combustion do not take place instantaneously. For example, at high speeds, combustion may begin when the crank is 30° or more before TDC rather than at the TDC position.
- The actual diagram has a pumping loop, which occurs because of pressure drops in the inlet manifold and exhaust system. These reduce the intake pressure to below atmospheric and increase the exhaust pressure to above atmospheric.
- Heat flow from the cylinder occurs during both compression and expansion strokes so that these processes are polytropic rather than adiabatic.

As a result of these differences, the actual cycle has a considerably lower power output and efficiency than the theoretical cycle.

Advantages and disadvantages of the Otto cycle

Advantages

Otto-cycle engines are by far the most dominant heat engine in use in the world today mainly because they have two outstanding advantages:

- They are capable of running at high speeds and therefore have a high power output per unit mass (high specific power output). This is of particular importance in mobile applications such as motor cars and aircraft. For example, the Rolls Royce Merlin engine that powered the Spitfire fighter in World War II produced 1.5 MW of power from a 27 L engine, with a specific power output of 1.9 kW/kg.
- Because combustion is internal, the working substance is brought to high temperatures during combustion and this enables high efficiencies to be achieved (remember the Carnot limitation that upper-cycle temperature affects the maximum possible efficiency).

Disadvantages

There are two main disadvantages of Otto-cycle engines:

- They are relatively noisy and prone to vibration and may produce high levels of pollutants. However, advances in pollution-control technology and enforcement of tighter controls have resulted in considerably reduced pollutant levels from modern engines.
- They are not efficient at low-power outputs (low-load conditions) because load control is achieved by throttling the air or air–fuel mixture coming into the engine. This reduces the intake pressure, which reduces the pressure and temperature after compression and after combustion and hence reduces both the power output and the efficiency.

Theoretical work and efficiency for the Otto cycle

Compression and expansion are adiabatic, so heat flow occurs only during the constant-volume portions of the cycle.

Heat supplied, $Q_S = mc_v(T_3 - T_2)$
Heat rejected, $Q_R = mc_v(T_4 - T_1)$ (positive)
Work, $W = Q_S - Q_R$

Cycle efficiency, $\eta = \dfrac{\text{work}}{\text{heat supplied}} = \dfrac{W}{Q_S}$

Example 6.4

For the theoretical Otto cycle with compression ratio = 8:1, inlet conditions = 97.5 kPa (abs.) and 50°C, and heat supplied = 950 kJ/kg of air, determine:
(a) the maximum cycle temperature;
(b) the indicated work per kilogram of air used;
(c) the efficiency.
Assume air may be treated as a perfect gas with $c_P = 1.005$ kJ/kgK, $c_v = 0.718$ kJ/kgK.

Solution

Refer back to Figure 6.12

$p_1 = 97.5$ kPa
$T_1 = 50°C = 323$ K
$\dfrac{V_1}{V_2} = 8$ (compression ratio)
$Q_S = 950$ kJ/kg

(a) For adiabatic compression ① → ②:

$$\frac{T_2}{T_1} = \left(\frac{V_1}{V_2}\right)^{\gamma-1}$$

where $\gamma = \dfrac{c_p}{c_v} = 1.4$

$$\therefore T_2 = 323 \times 8^{0.4}$$
$$= 742 \text{ K}$$

For constant-volume heating, ② → ③:

$$Q_S = mc_v(T_3 - T_2)$$
$$950 = 1 \times 0.718 \times (T_3 - 742).$$
$$\therefore T_3 = \textbf{2065 K (1792°C)} \text{ (maximum cycle temperature)}$$

(b) For adiabatic expansion ③ → ④:

$$\frac{T_4}{T_3} = \left(\frac{V_3}{V_4}\right)^{\gamma-1}$$

$$\therefore T_4 = 2065 \times \left(\frac{1}{8}\right)^{0.4}$$

$$\therefore T_4 = 899 \text{ K}$$

For constant-volume heat rejection ④ → ①:

$$Q_R = mc_v(T_4 - T_1)$$
$$= 1 \times 0.718 \times (899 - 323)$$
$$= 413.5 \text{ kJ/kg}$$
$$W = Q_S - Q_R$$
$$= 950 - 413.5$$
$$= \textbf{536.5 kJ/kg}$$

(c) $\eta = \dfrac{W}{Q_S}$

$$= \frac{536.5}{950} \times 100$$
$$= \textbf{56.5\%}$$

Notes

1 The work output could also have been calculated from the area under the adiabatic expansion and compression curves.

2 The maximum theoretical efficiency of a Carnot-cycle heat engine, working between the same temperature limits, is

$$\eta_C = 1 - \frac{T_c}{T_h} = 1 - \frac{323}{2065} = 84.4\%$$

Hence the Otto cycle has considerable lower efficiency than the Carnot cycle for the same temperature limits.

Effect of compression ratio on work output and efficiency of the Otto cycle

The theoretical efficiency of the Otto cycle can be expressed in terms of compression ratio V_1/V_2 by the following equation (derived in Appendix 12):

$$\eta = 1 - \frac{1}{\left(\dfrac{V_1}{V_2}\right)^{\gamma-1}}$$

theoretical efficiency of Otto cycle (6.6)

Hence the efficiency of the ideal Otto cycle depends only on the compression ratio. The higher the compression ratio, the higher the efficiency and therefore the work output from the cycle.

Example 6.5

Obtain the efficiency of the theoretical Otto-cycle engine given in Example 6.4 using Equation 6.6.

Solution

Compression ratio $V_1/V_2 = 8$, $\quad \gamma = 1.4$

$$\eta = 1 - \frac{1}{\left(\dfrac{V_1}{V_2}\right)^{\gamma-1}}$$

$$= 1 - \frac{1}{8^{0.4}}$$

$$= \mathbf{0.565} \text{ or } \mathbf{56.5\%}$$

 ## *Self-test problem 6.5*

For the theoretical Otto cycle, draw a graph showing temperature after compression T_2, and efficiency η versus compression ratio for compression ratios of 1, 2, 3, 4, 5, 7.5, 10, 15, and 20 : 1. Assume $T_1 = 15°C$, $\gamma = 1.4$. Comment on the result.

6.9 DIESEL CYCLE

The Diesel cycle is a positive-displacement, four-stroke, internal-combustion, compression-ignition, open cycle invented by Rudolf Diesel in the 1860s. The sequence of processes in this cycle is shown in the theoretical *p–V* diagram in Figure 6.13 on page 140.

When compared with the theoretical Otto cycle (Fig. 6.12 on page 136), the similarities are evident. In fact, the only point of difference is that after adiabatic compression, combustion takes place at constant pressure rather than at constant volume. The actual *p–V* diagram differs from the theoretical in the same way as the actual Otto cycle differs from the theoretical, namely by having rounded rather than sharp corners, by the pumping loop and because heat losses occur during the actual compression and expansion processes.

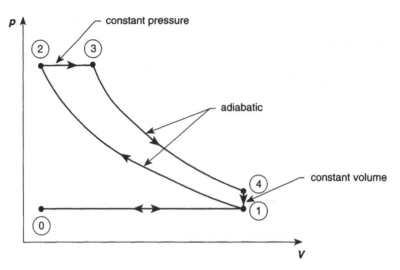

Fig. 6.13   *Theoretical Diesel cycle*

Diesel-cycle engines differ from Otto-cycle engines in the following ways:

- Air only is drawn into the cylinder and compressed (not an air–fuel mixture).
- Compression ratios above 12 : 1 are used, 16 : 1 to 18 : 1 being usual. This raises the air to a temperature high enough to ensure spontaneous combustion. In spark-ignition engines compression ratios of less than 10 : 1 are usual.
- Diesel engines do not have a spark plug but have an injector in each cylinder, which injects the fuel directly into the cylinder at (or near) the top of the stroke. Spark-ignition engines may also use fuel injectors rather than carburettors but, in such cases, the fuel is injected through the inlet valve port and the air–fuel mixture is compressed. Diesel engines compress air only.
- Injection of the fuel (and combustion of it) continues into the power stroke and this tends to maintain a constant pressure while injection (and combustion) takes place.
- Diesel-engine-load control is achieved by varying the amount of fuel injected and not by throttling the air inlet. Spark-ignition engines require a constant air–fuel mixture to maintain combustion under all conditions of speed and load but this is not the case with diesel engines. Under low-load conditions, diesel engines run with a lean mixture (excess air). The mixture becomes progressively richer as the load increases and more fuel is injected during each power stroke.

Advantages and disadvantages of the Diesel cycle

Advantages

Next to Otto-cycle engines, diesel engines are the most common type of heat engine in use in the world today. They have several important advantages:

- Diesel engines are the most efficient of all the heat engines. This follows from the fact that they are able to use high compression ratios and because they are able to use excess air to ensure thorough combustion of the fuel.
- Because diesel engines do not require electrical equipment to operate and do not need a precise air–fuel ratio for combustion, they are very reliable and will operate under adverse conditions. For example, moisture tends to cause malfunction of the high-

tension electrical gear in spark-ignition engines but this is not a problem in diesel engines.

- Diesel engines use excess air and do not require fuel additives and therefore produce less combustion pollutants. Also, the fuel is less volatile and less fuel vaporisation to the atmosphere occurs.
- Diesels engines are the safest of all the heat engines. Diesel fuel is less volatile than petrol and does not have the explosive problems associated with gaseous fuels. Therefore diesels are used in preference to other types of heat engines in applications such as marine installations, where the engine is located in a confined space and where fire or explosion risk is a major consideration.

Disadvantages

Diesel engines have two main disadvantages:

- They have lower specific power output than spark-ignition engines. This is mainly because combustion is slower and this limits the operating speed. Lower speed means lower power output because of the basic relationship between power and speed, $P = T\omega$. Also the higher compression ratio requires a heavier flywheel and starting equipment. For example, compare a Rolls Royce compression-ignition engine of 26 L capacity, producing 900 kW and with peak specific power output of 0.5 kW/kg, with the Merlin spark-ignition engine, which had a peak specific power output of 1.9 kW/kg.
- They cost more than spark-ignition engines. This is because of the greater mass of the engine (more materials) and the higher cost of fuel-injection and starting equipment.

Theoretical work and efficiency of the Diesel cycle

The theoretical work and efficiency of the Diesel cycle may be calculated in the same manner as for the Otto cycle except that the heat input is given by:

$$Q = mc_p(T_2 - T_1)$$

because heat flow occurs at constant pressure and not constant volume.

Note   Equation 6.6 applies only to the Otto cycle and not the Diesel cycle.

Self-test problem 6.6

Repeat the calculations of Example 6.4 for a theoretical Diesel cycle with compression ratio 20:1.

6.10 *DUAL CYCLE*

The assumption of either constant-volume *or* constant-pressure combustion is an oversimplification; neither occurs in practice. However, a more accurate representation of the combustion process in high-speed internal-combustion engines can be obtained by assuming there is an initial sharp pressure increase (at constant volume) followed by essentially constant-pressure combustion.

The theoretical cycle based on a combination of constant-volume and constant-pressure combustion is called the **dual cycle** or **mixed cycle** (Fig. 6.14 on page 142). Calculation follows the same method of stepwise movement around the cycle.

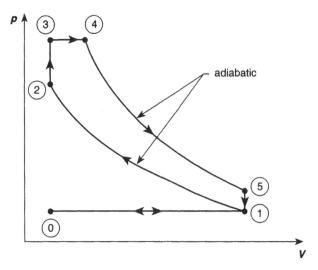

Fig. 6.14   *Theoretical dual (or mixed) cycle*

6.11 *TWO-STROKE ENGINES*

There is no unique two-stroke cycle, but the principle of operation may be applied to the Otto, Diesel or dual cycles. In the ideal case, the only difference between the p–V diagrams for two-stroke and four-stroke operation is the elimination of the constant-pressure induction stroke ⓪ → ① and the constant-pressure exhaust stroke ① → ⓪. In theory, then, there is the same indicated work output per kilogram of working substance and the same indicated thermal efficiency for the cycle in both two-stroke and four-stroke operation. However, for the same-capacity engine, two-stroke operation gives theoretically twice the indicated power output since there is double the flow rate of working substance through the engine at any given speed. In practice, this advantage may not be completely realised, particularly with spark-ignition engines, where the two-stroke cycle has a lower efficiency than the corresponding four-stroke cycle.

The two-stroke spark-ignition engine

Engines constructed to operate on this cycle do not require inlet and exhaust valves but use the piston to open and close ports at the bottom end of the cylinder. Also, the underside of the piston is used to slightly compress the charge which has been drawn into the crankcase. The piston itself may have a flat or raised head according to the port design. The basic principle of operation is made clear in Figure 6.15.

Of utmost importance to the efficient operation of the engine is **scavenging**, that is, replacing the products of combustion with fresh charge. Even with careful port and piston design and tuning of exhaust systems, it is difficult to scavenge as effectively as with four-stroke engines. This is the reason why two-stroke spark-ignition engines do not achieve double the power output with the same efficiency as equal-capacity four-stroke engines.

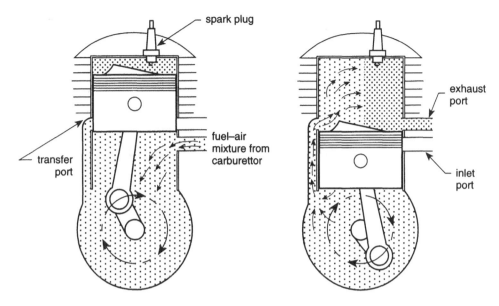

Fig. 6.15   *Two-stroke spark-ignition engine*

Another difficulty is that the induction of fuel mixture into the crankcase would soon dilute the lubricating oil if a normal wet-sump system of lubrication were used. One solution to this problem is to dispense with the oil sump and oil pump and mix the lubricating oil with the fuel in proportions such as 25:1, 50:1 or 100:1. A recent innovation is an oil-injection system in which lubricating oil is injected rather than being mixed with the fuel. However, with either system, lubricating oil is burnt and appears as smoke in the exhaust.

A further disadvantage of these engines is that the effectiveness of scavenging is very sensitive to exhaust back-pressure brought about by the silencing equipment, and so they are noisier because less effective silencers are used.

As a result of these disadvantages, two-stroke spark-ignition engines are not used for motor-vehicle engines, particularly with the continuing drive for reduction in exhaust noise and emissions. They are, however, still used for small power units such as lawn-mower engines, because of their simplicity and low cost. They are also used for engines that require a high specific power output such as outboard motors for use with power craft.

The two-stroke compression-ignition engine

The general arrangement of this engine is shown in Figure 6.16 on page 144.

A blower provides efficient scavenging since a surplus of air effectively purges the exhaust gas. (This cannot be done with two-stroke spark-ignition engines because unburnt fuel would also be blown out the exhaust.) This arrangement allows normal wet-sump lubrication and combines the advantages of compression-ignition with two-stroke operation.

Engines of this type are used where high power output is required, for example in the engines used to power large ships. In such cases, exhaust turbines are used to power the intake-air blowers and this enables high overall efficiencies (of the order of 50%) to be achieved. Such engines are very often built in the double-acting configuration and can be designed to work in either direction, thus eliminating the need for reversing gear.

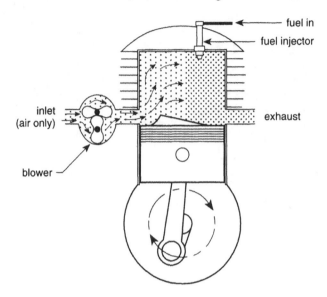

Fig. 6.16   *Two-stroke compression-ignition engine*

6.12 *JOULE CYCLE*

This cycle is used with gas turbines, jet engines and steam turbines. With gas turbines and jet engines, it is an open, internal-combustion cycle as illustrated in Figure 6.17 (opposite).

The cycle may also be made a closed, external-combustion cycle by inclusion of the return path ④ → ①. In this form it is used for steam turbines, where the return path is provided by the condenser, which condenses the steam to water. A boiler feed pump is used (rather than a compressor) to pump the water into the boiler, where steam is formed to drive the turbine.

Advantages and disadvantages of the Joule cycle

As a general principle, gas and steam turbines are more costly and less efficient than comparable diesel engines and are used in preference to diesel engines only in specific applications. One such application is that of large-scale electricity production, where steam turbines are used. Here, efficiency-improving devices (such as air pre-heaters and economisers) can be incorporated to improve the efficiency. Also, because combustion is external, coal can be used for a fuel and this is a far cheaper and more plentiful fuel than petroleum derivatives. Jet engines find a niche as propulsion units for large aircraft because the high energy associated with the high flow rate of the hot exhaust gas provides forward propulsive thrust, which considerably improves the efficiency. Also these engines have a very high specific power output and this is of critical importance in aircraft applications.

It is interesting to note that the Joule cycle may also be used in reverse as a refrigeration cycle. Here, shaft power is input and the air is cooled after compression to around atmospheric temperature. After expansion in the turbine, the air temperature drops to below atmospheric and may be used to provide refrigeration. This cycle (air-cycle refrigeration) is used for aircraft refrigeration and air conditioning.

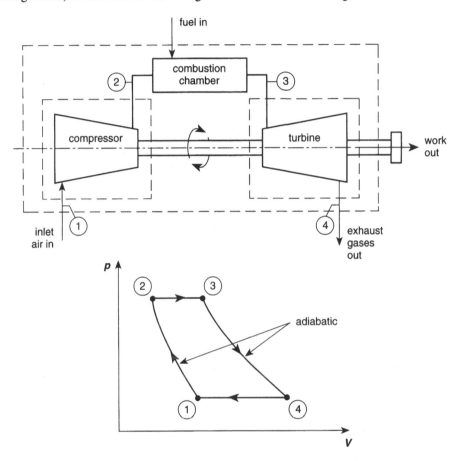

Fig. 6.17   *Joule cycle (open)*

Summary

A heat engine is a mechanical device that converts heat to work continuously. The fundamentals of a heat engine are that it must have a heat source, heat sink, working substance, working cycle and mechanical work or power output. Unlike other energy-conversion devices that may have a theoretical or actual efficiency approaching 100%, the efficiency of a heat engine is limited to well below 100% and depends on the maximum and minimum temperatures at which heat is transferred into and out of the working substance. The limiting efficiency is known also as the Carnot efficiency and a heat-engine cycle with this efficiency is known as the Carnot cycle. This cycle consists of two isothermal and two adiabatic processes but it is not a practical cycle.

A practical cycle with the same efficiency as the Carnot cycle is the Stirling cycle, with regenerative heat transfers. Stirling engines have many advantages, which stem mainly from the fact that the cycle is a closed one with external combustion. However, these engines also have many disadvantages, which have precluded their widespread usage.

For high specific power output and relatively high efficiency, four-stroke spark-ignition engines operating on the Otto cycle have many applications such as engines for motor vehicles. The work output and efficiency of the cycle depend primarily on the compression ratio but there is a limit to the compression ratio that can be used because detonation (or pinging) occurs at high compression ratios. This disadvantage is overcome in the Diesel cycle because only air is compressed and not an air–fuel mixture. This enables high compression ratios to be used and gives these engines high efficiency.

Two-stroke operation enables high specific power outputs because there are twice as many working strokes compared with four-stroke operation. Two-stroke petrol engines are not widely used but two-stroke diesel engines are used when large amounts of power are needed. Sometimes a double-acting configuration is used.

The Joule cycle is applicable to gas turbines, jet engines and steam turbines. Steam turbines are used mainly for large-scale power generation and jet engines find application as power plants for large aircraft.

 Problems

Note For these problems, assume ideal gas conditions and ideal processes. Use the values: $c_p = 1.005$ kJ/kgK, and $c_v = 0.718$ kJ/kgK for air.

6.1 Giving reasons for your answer, determine which of the following are heat engines and therefore limited by Carnot efficiency:

(a) a cannon, using combustion of an explosive to fire the shell;

(b) a wind turbine (windmill);

(c) an electric motor;

(d) a solar hot-water heater;

(e) a solar-powered Stirling engine;

(f) a fuel cell (which converts hydrogen and oxygen directly into electricity) coupled to an electric motor;

(g) the braking mechanism of a motor vehicle;

(h) an exhaust-gas supercharger;

(i) an air pre-heater in which air is pre-heated by exhaust gas;

(j) a bimetal rod, which rotates when exposed to sunlight;

(k) a device that converts wave energy into mechanical energy.

6.2 An internal-combustion engine has a minimum combustion temperature of 1500°C. The exhaust-gas temperature is 450°C. What is the maximum possible efficiency of the engine?

When an exhaust-gas supercharger is fitted, the exhaust-gas temperature drops to 250°C. What effect does this have on the maximum possible efficiency?

59.2%; increases to 70.5%

6.3 Briefly discuss the essentials of a heat engine.

A theoretical Carnot-cycle engine has an efficiency of 60% and operates at a minimum temperature of 20°C. Determine the maximum cycle temperature and the rate of heat supply for a power output of 24 kW.

460°C; 40 kW

6.4 An internal-combustion engine uses fuel, of energy content 44.4 MJ/kg, at a rate of 5 kg/h. If the efficiency is 28%, determine the power output and the rate of heat rejection.

17.3 kW; 44.4 kW

6.5 A heat engine operates with a source temperature of 1000 K and a sink temperature of 300 K. Which of the following is the most effective way of improving the performance of the engine?

(a) Increase the source temperature to 1100 K.

(b) Decrease the sink temperature to 200 K.

(b) is much better.

6.6 An oil-fired steam generator is connected to a steam engine, which produces 50 kW of power when using oil of energy content 47 MJ/kg at a rate of 19.5 kg/h. Steam temperature is 500°C and cooling-water temperature is 15°C. Calculate:

(a) the heat-rejection rate (MJ/h);

(b) the overall efficiency of the plant;

(c) the maximum efficiency attainable by the engine.

(a) 736.5 MJ/h   (b) 19.6%   (c) 62.7%

6.7 An engineer has designed an engine that, on test, gave the following results: heat supplied = 25 kJ/s at 600°C; heat rejected = 12 kJ/s at 200°C; work output = 15 kW. Analyse the feasibility of this engine, with respect to:

(a) energy balance;

(b) Carnot efficiency.

Impossible result in both respects

6.8 A comparison of two types of heat engines, using different sources of heat, is given in the following table:

	Type A	Type B
Source temperature (°C)	1000	700
Power output (kW)	40	50
Heat-supply rate (kW)	95	125

If both types use the atmosphere at 20°C for a heat sink, determine which engine type:

(a) is more efficient;

(b) utilises the available temperature of the heat source most efficiently (that is, has the higher ratio of efficiency to maximum possible efficiency).

(a) Type A   (b) Type B

6.9 A Stirling-cycle engine works between temperature limits of 550°C and 20°C. What would be the heat-input rate per kilowatt of power produced and the overall efficiency of the engine if it has an actual efficiency of 50% of ideal?

3.11 kW; 32.2%

6.10 Draw the Carnot cycle on a p–V diagram. For such a cycle using air as the working substance, the maximum and minimum pressures during adiabatic compression are 900 kPa and 300 kPa respectively. If the minimum temperature is 40°C, determine the theoretical maximum temperature and efficiency.

155°C; 26.9%

6.11 An ocean thermal energy collector is to produce 1 MW of power and operate in a location where the surface temperature of the sea is 24°C and the deep temperature is 8°C. The working fluid will achieve a maximum temperature of 21°C and a minimum temperature of 11°C. The overall efficiency of the plant is expected to be 50% of the ideal maximum efficiency for the same temperature difference of the working fluid. The specific heat capacity of sea water may be taken as 4 kJ/kgK. Determine:
(a) the efficiency of the plant;
(b) the heat-supply rate;
(c) the flow rate of hot sea water;
(d) the heat-rejection rate;
(e) the flow rate of cold sea water.

(a) 1.7% (b) 58.8 MW (c) 4.9 m³/s (d) 57.8 MW (e) 4.82 m³/s

6.12 Repeat the calculations for Self-test problem 6.4 for an actual engine having an efficiency of 40% of the ideal.

(a) 21.5% (b) 679 W (c) 146 W (d) 533 W

6.13 Give a common practical application for each of the following engine types, indicating why the particular type has been selected in each case:
(a) four-stroke spark-ignition engine;
(b) two-stroke spark-ignition engine;
(c) four-stroke compression-ignition engine;
(d) two-stroke compression-ignition engine;
(e) fan-jet engine.

6.14 Briefly compare and contrast spark-ignition and compression-ignition engines (giving reasons) under the following headings:
(a) cost of manufacture (for given power output);
(b) specific power (power per unit mass);
(c) efficiency;
(d) reliability;
(e) safety;
(f) pollutants.

6.15 Describe briefly *three* advantages and *three* disadvantages of Stirling-cycle engines compared with Otto-cycle engines.

6.16 Draw the constant-volume (Otto) cycle on a p–V diagram. For a theoretical engine operating on this cycle, drawing in air at 20°C and 100 kPa, and operating on a compression ratio of 8:1, determine:
(a) the temperature and pressure after compression;
(b) the work done during the compression stroke if the swept volume of the cylinder is 0.5 L.

(a) 1.84 MPa; 400°C (b) –185 J

6.17 If the engine in Problem 6.16 burns fuel so that the pressure after combustion is 1.8 times the pressure before combustion, determine:

(a) the temperature after combustion;

(b) the work done during the power stroke;

(c) the nett work of a complete cycle.

(a) 938°C   (b) 333 J   (c) 148 J

6.18 Draw the constant-pressure (Diesel) cycle on a p–V diagram. For a theoretical engine operating on this cycle, drawing in air at 20°C and 100 kPa, and operating on a compression ratio of 18:1, determine:

(a) the temperature and pressure after compression;

(b) the work done during the compression stroke if the swept volume of the cylinder is 0.5 L.

(a) 5.72 MPa; 658°C   (b) –288 J

6.19 If the engine in Problem 6.18 burns fuel so that the volume after combustion is twice the volume before combustion, determine:

(a) the temperature after combustion;

(b) the work done during the power stroke;

(c) the nett work of a complete cycle.

(a) 1589°C   (b) 660 J   (c) 372 J

6.20 For a theoretical Otto cycle operating on a compression ratio of 8.5:1, taking in air at 15°C and 101.3 kPa and having a maximum cycle temperature of 1800°C, calculate:

(a) the temperature after compression (T_2);

(b) the temperature after expansion (T_4);

(c) the theoretical work of compression per kg of air;

(d) the theoretical work of expansion per kg of air;

(e) the nett work per kg of air;

(f) the heat supplied per kg of air;

(g) the theoretical cycle efficiency using data previously calculated;

(h) the theoretical cycle efficiency using Equation 6.6.

(a) 404.9°C   (b) 607.7°C   (c) –279.8 kJ   (d) 855.5 kJ   (e) 575.7 kJ   (f) 1002 kJ

(g) 57.5%   (h) 57.5%

6.21 An engine operates on the dual cycle with compression ratio of 16:1, and takes in air at atmospheric pressure and at a temperature of 21°C. The heat energy supplied is 430 kJ/kg, of which 35% is supplied at constant volume and the remainder at constant pressure. Calculate the following:

(a) the temperature after compression (T_2);

(b) the maximum cycle temperature (T_4)

(c) the maximum cycle pressure (p_3 or p_4)

(a) 618°C   (b) 1106°C   (c) 6.07 MPa

6.22 An engine working on the Otto cycle has a theoretical cycle efficiency of 70% of that of a Carnot-cycle engine working between the same upper- and lower-cycle temperatures of 1550°C and 20°C respectively. Determine the compression ratio of the engine.

9.15 : 1

CHAPTER 7

Heat-engine performance ≡

Objectives

On completion of this chapter you should be able to:

- describe the common heat-engine-performance criteria, namely torque, brake power, indicated power, friction power, heat-supply rate (energy-input rate), fuel consumption, specific fuel consumption, efficiency (overall), mechanical efficiency, indicated thermal efficiency and volumetric efficiency;
- calculate performance criteria from typical test data;
- explain how the various performance criteria can be measured; for brake power describe the various braking systems such as rope brake, shoe brake, hydraulic dynamometer and electric dynamometer; for indicated power, describe the principles of operation of a mechanical indicator, electrical or electronic indicator and the Morse test;
- balance the energy inputs and outputs for a heat engine from typical test data; divide the output energy into typical categories such as power, energy loss to cooling, energy loss to exhaust and other energy losses;
- describe two common heat-engine-performance tests, namely variable speed and load at constant-throttle setting and variable load and throttle setting at constant speed;
- interpret performance curves for a heat engine;
- draw performance curves for a heat engine given appropriate test data.

Introduction

In Chapter 6, heat engines were treated from an essentially theoretical viewpoint. In this chapter, the practical aspects of heat-engine performance are examined. The important criteria are defined and calculated from typical test data. Various testing methods are described and the results of these tests shown as performance curves.

7.1 *POWER OUTPUT*

The power output (*P*) of a heat engine is also known as the **brake power** because it may be measured by applying a brake to the engine to absorb the power. It is also known as **developed power** or **shaft power.**[*]

Most heat engines produce rotational power at the output shaft, given by Equation 3.9:

$$P = T\omega$$

where P = power output (W)
  T = output or brake torque (Nm)
  ω = angular speed of the shaft (rad/s)

$$= \frac{2\pi N}{60}$$

$$= \frac{\pi N}{30}$$

where N = rotational shaft speed (rpm)

Measurement of power output

The power output of an engine may be measured by a number of methods, all of which consist of applying a **resisting torque**. The resisting torque is measured from the force applied and the lever-arm length or distance from the force to the axis of rotation. At the same time, the speed may be measured by a tachometer (mechanical or electrical).

The rope brake

One of the earliest methods of measuring the power output of an engine was to wind a rope around the flywheel, and attach the upper end to a fixed-spring balance and the lower end to a weight pan, as shown in Figure 7.1 (on page 152).

It is necessary to provide cooling to the flywheel rim such as a water-drip (not shown) in order to prevent the rope from getting too hot. Even so, the rope brake is only suitable for low-power, slow-speed engines with large flywheels.

If the system is in a steady-state of operation the output torque is:

$$T = (F_2 - F_1)r$$

where F_2 = the weight on the rope (N)
  F_1 = the spring-balance reading (N)

$$r = \frac{\text{diameter of flywheel} + \text{diameter of the rope}}{2} \text{ m}$$

and the output power is:

$$P = T\omega$$

---

[*] Power output is known in the imperial system as 'horsepower' because the basic unit is a measure of the average power output of a horse.
  The conversion is: 747 W = 1 Hp or 1 kW = 1.34 Hp.

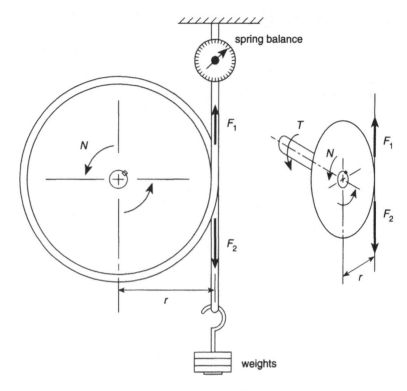

Fig. 7.1   *A rope brake*

Notes

1 The resisting torque is due to friction between the rope and the flywheel but the coefficient of friction and the number of turns of rope around the flywheel do not affect the calculation.

2 The spring balance would not be necessary if the weight were adjusted to give a zero slack side tension in the top rope. This is impractical and it is more convenient to adjust the weight to give a convenient spring-balance reading.

3 The weight of the rope is usually negligible and does not affect the result.

4 The spring-balance reading may fluctuate (particularly if the engine is a single-cylinder four-stroke) due to the torque fluctuations during the cycle. It is necessary to obtain an average reading.

 ## Self-test problem 7.1

For a rope brake attached to an engine flywheel, the following data were recorded:

> diameter of flywheel = 1.35 m
> diameter of rope = 25 mm
> flywheel speed = 225 rpm
> average spring-balance reading = 9 N
> applied weight = eight 5 kg masses

Determine the power output of the engine. What would be the effect of adding one extra 5 kg mass, or taking away one 5 kg mass?

The shoe brake

A typical shoe brake is illustrated in Figure 7.2. Here the braking torque is applied by brake shoes similar to those used on motor-vehicle drum brakes but a disc-brake system could also be used. Braking torque may be applied by tightening the tie rods and measured by measuring the force applied to a lever arm fitted with a suitable force-measuring device. The simplest such device is a spring balance (provided it has a suitable force-range scale) or a weight–spring-balance system (as illustrated with the rope brake in Fig. 7.1). The arm needs to be provided with limit stops to limit the amount of movement. Some form of auxiliary cooling may be necessary on the brake shoes to prevent them from getting too hot.

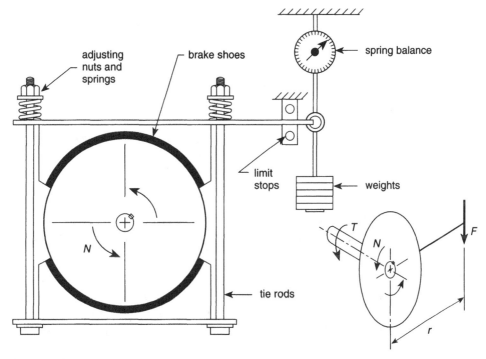

Fig. 7.2   *A shoe brake*

The output torque is:

$$T = Fr$$

where F is the spring-balance reading if a spring balance is used alone or the difference between the applied weights and the spring-balance reading if a weight–spring-balance system is used.

The power output is as before:

$$P = T\omega$$

The hydraulic dynamometer

In the rope brake and shoe brake, mechanical energy is converted to heat energy by mechanical (dry) friction. With high-power-output engines it is difficult to dissipate the heat effectively by this method and they tend to become very hot. This disadvantage is overcome in the hydraulic dynamometer because fluid friction is used rather than mechanical friction. The principle is illustrated in Figure 7.3 on page 154.

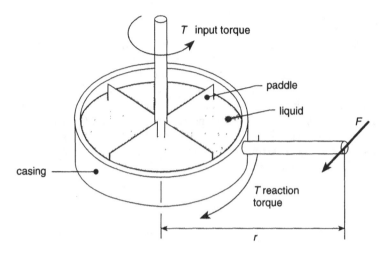

Fig. 7.3   *Principle of the hydraulic dynamometer*

In practice, in order to increase fluid friction and hence the torque and power capability, the casing is fitted with vanes and the paddle blades are shaped so that the whole device becomes in effect a fluid coupling (as used in automatic car transmissions).

The output torque and power are calculated from the force and speed measurements as before by:

$$T = Fr \text{ and } P = T\omega$$

The electric dynamometer

One form of electric dynamometer is illustrated in Figure 7.4 (opposite). Essentially it is an electric generator (or dynamo) in which the majority of the output power of the engine is converted into electrical energy and then dissipated as heat in load resistors. Some of the output energy is also dissipated in internal-resistance losses and air friction.

In order to measure the output torque it is necessary to mount the casing of the generator on rollers or similar devices so that it can swing freely (through a restricted arc of movement). An arm is attached to the casing and a force-measurement system attached to the arm (as before).

A variant of this type is the **eddy-current dynamometer** in which all the power is dissipated in the generation of eddy currents, and heat is carried away by a flow of cooling water.

The output torque and power are calculated from the force and speed measurements as before by:

$$T = Fr \text{ and } P = T\omega$$

Example 7.1

A heat engine is coupled to an electric dynamometer whose electrical current output is dissipated in a bank of water-cooled resistors. The following data are recorded:

> length of load arm = 960 mm
> spring-balance reading = 15 N
> applied weight = five 10 kg masses
> rotational speed = 1500 rpm
> cooling-water flow rate = 180 L/min
> temperature rise of the cooling water = 5.2°C

Determine:

(a) the brake power;

(b) the power dissipated in electric-generator losses.

Solution

(a) $P = T\omega$

$$= [(5 \times 10 \times 9.81) - 15] \times 0.96 \times \frac{\pi \times 1500}{30} \quad \text{W}$$

$$= \textbf{71.7 kW}$$

(b) The power dissipated by the cooling water is the heat-flow rate $\dot{Q}$

$$\dot{Q} = \dot{m}c\,(T_2 - T_1)$$

$$= \frac{180}{60} \times 4.19 \times 5.2$$

$$= 65.4 \text{ kW}$$

Hence the power dissipated in electric-generator losses is

$$71.7 - 65.4 = \textbf{6.34 kW}$$

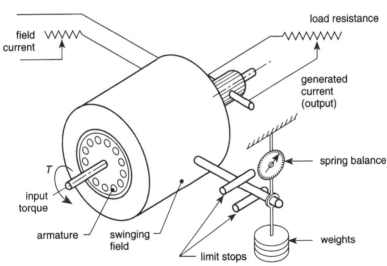

Fig. 7.4   *Electric dynamometer*

7.2 *HEAT-SUPPLY RATE*

The heat-supply rate ($\dot{Q}_s$) is the energy, in the form of heat, supplied per second to the engine.

The heat-supply rate from the combustion of a fuel is given by:

$$\boxed{\dot{Q}_s = \dot{m}E}$$

heat-supply rate (7.1)

where $\dot{Q}_s$ = heat-supply rate (MJ/s or MW)

   $\dot{m}$ = mass-flow rate (kg/s)

   E = energy content of the fuel (MJ/kg)

The energy content of the fuel needs to be known or determined beforehand using a bomb calorimeter (solid or liquid fuels) or a gas calorimeter (gaseous fuels). The mass-flow rate of the fuel is usually difficult to measure directly but may be determined by measuring the volume-flow rate and density of the fuel and multiplying them together. The volume-flow rate may be measured by diverting the fuel to the engine through a bulb of known volume and measuring the time taken for the fuel to pass through the bulb. The density of the fuel may be measured by accurately weighing a known volume of fuel.

7.3 *EFFICIENCY*

The purpose of a heat engine is to convert heat energy into mechanical energy. The efficiency (η) (known also as thermal efficiency, brake thermal efficiency or overall efficiency) is given by Equation 6.3:

$$\eta = \frac{W}{Q_s} = \frac{P}{\dot{Q}_s}$$

Self-test problem 7.2

A four-stroke, spark-ignition engine is coupled to a hydraulic dynamometer. From the following test results determine the power output, heat-supply rate and efficiency of the engine:

> energy content of the fuel = 45 MJ/kg
> dynamometer load arm length = 480 mm
> engine speed = 3500 rpm
> relative density of the fuel = 0.8
> dynamometer force reading = 200 N
> time to consume 50 mL of fuel = 15 s

7.4 *SPECIFIC FUEL CONSUMPTION*

When comparing engines that use the same type of fuel (for example spark-ignition engines using petrol or compression-ignition engines using dieselene), it is often useful to calculate the specific fuel consumption (SFC), which is the quantity of fuel consumed per unit of energy output of the engine. The unit of energy output of the engine may be in megajoules or kilowatt hours, where

> 1 kWh = 1 kW for 1 h
> = 1000 W × 3600 s
> = 3.6 MJ.

For example, a specific fuel consumption of 0.4 kg/kWh = 0.111 kg/MJ.

Relationship between specific fuel consumption and efficiency

Specific fuel consumption and efficiency measure the same thing but in different ways. For example, the best efficiency usually attained by a petrol engine is about 35%. Petrol has an average energy content of about 46.5 MJ/kg, so the specific fuel consumption would be 1/46.5 kg/MJ if the engine were 100% efficient. Since the engine is 35% efficient, the specific fuel consumption at maximum efficiency is:

$$\frac{1}{46.5 \times 0.35} = 0.06 \text{ kg/MJ}$$
$$= 0.06 \times 3.6 \text{ kg/kWh}$$
$$= 0.22 \text{ kg/kWh}$$

This is a useful figure because it can now be used to calculate the fuel consumption of any size of equivalent engine. For example, a 50 kW petrol engine at 35% maximum efficiency could be expected to consume about $50 \times 0.22 = 11$ kg of petrol per hour. Petrol has an average relative density of about 0.75, so the fuel consumption in litres per hour is:

11/0.75=14.7 L/h

Similarly, for a diesel engine with efficiency 40%, using dieselene with energy content 45.6 MJ/kg and relative density 0.85, the specific fuel consumption is:

0.055 kg/MJ = 0.2 kg/kWh
$$= 0.23 \text{ L/kWh}$$

 ## *Self-test problem 7.3*

A diesel engine consumes heavy fuel oil that has an energy content of 42.6 MJ/kg and relative density 0.9. When producing 200 kW shaft power, the engine consumes 3.85 L of fuel over a 5-min test period. Determine:
(a) the efficiency;
(b) the specific fuel consumption in kg/kWh.

7.5 *INDICATED POWER*

Indicated power (IP) is the power developed by the working substance. Since the working substance is usually a fluid, indicated power is also **fluid power**.

Indicated power derives its name from the fact that one of the earliest methods of determining it was by means of an engine indicator (invented by James Watt).

A heat engine operates in a cycle, so the indicator power is the indicated work per cycle multiplied by the number of cycles per second:

$$IP = W_{\text{ind}}n$$

where *IP* is in watts if *W* is in joules and *n* is in cycles per second.

Measurement of indicated work

Indicated work may be determined experimentally by one of three different methods, namely mechanical, electrical or electronic methods.

The mechanical indicator

The operating principle of a mechanical-engine indicator is illustrated in Figure 7.5(a). Pressure changes in the cylinder cause the slave piston to move so that the stylus traces out a line that is in fact a *p–V* diagram to some scale because the vertical movement is proportional to pressure and the horizontal movement is proportional to volume. The practical adaptation of this principle is illustrated in Figure 7.5(b) (opposite).

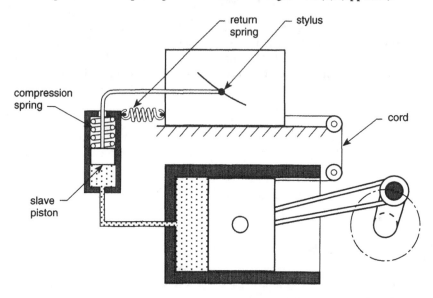

Fig. 7.5 (a) *Mechanical-indicator operating principle*

Due to the effects of inertia and friction, mechanical indicators are suitable only for slow speeds (less than 1000 rpm).

The electrical indicator

A typical electrical indicator is the Farnboro Indicator, which has an externally pressurised diaphragm connected to the engine cylinder. When cylinder pressure slightly exceeds the external pressure, electrical contact is made and a spark is discharged through a stylus onto paper fitted to a revolving drum attached through a clutch mechanism to the crankshaft. As the external pressure is changed, a new spark point occurs, and so, over a large number of engine cycles, a complete pressure trace is obtained.

Because the drum rotates and does not reciprocate and because the diaphragm is light and has only a small deflection, this indicator does not have the inertial effects of a mechanical indicator and is therefore suitable for higher engine speeds (up to about 4000 rpm). A disadvantage is that the diagram obtained is a pressure–crank-angle diagram (not a pressure–volume diagram) and conversion is necessary using drawing-construction or computer methods.

The electronic indicator

Electronic indicators use a pressure transducer connected to the cylinder and the pressure change is converted to a voltage change, which can then be displayed on an oscilloscope. Inertia and friction are entirely eliminated so these indictors are suitable for high-speed engines. Like the electric indicator, a *p–V* diagram is not obtained, rather a pressure–time diagram is displayed, which needs to be converted.

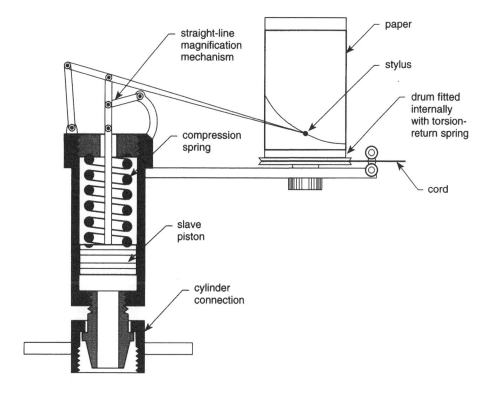

Fig. 7.5 (b)   *Practical adaptation of the mechanical engine indicator*

Calculating the indicated power

In order to calculate the indicated power, the indicated work needs to be determined from the indicator diagram. The work done is the area of the diagram (to some scale). The area may be determined by use of a planimeter (analog or digital). For four-stroke engines, the area of the pumping loop needs to be deducted from the area of the power loop in order to obtain the nett work per cycle. For multicylinder engines, each cylinder should be tested separately and the average (or mean) value obtained. However, if the engine is in good condition, there should not be a great deal of variation between the cylinders and in such cases it is often sufficiently accurate to test one cylinder and assume the value obtained is typical. If the engine is double acting, the indicated work on each side of the piston needs to be measured and the two added together for that cylinder.

For convenience of calculation, the area of the diagram is usually converted to a mean effective pressure (p_m) defined as the *constant pressure that, acting on the piston for one working stroke, would give the same indicated work as that actually obtained*. This concept is illustrated in Figure 7.6 on page 160.

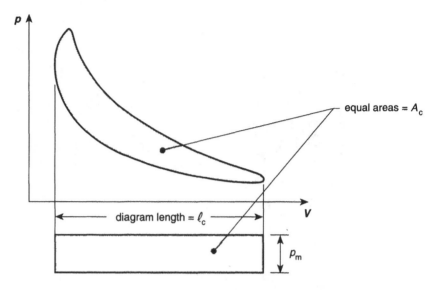

Fig. 7.6 *Using mean effective pressure, the shape of any indicator card is reduced to a rectangle*

From a mechanical-indicator diagram, the mean effective pressure may be calculated from the following equation:

$$p_{\mathrm{m}} = \frac{A_{\mathrm{c}} k_{\mathrm{s}}}{l_{\mathrm{c}}}$$ mean effective pressure (7.2)

where p_{m} = mean effective pressure (kPa)
   A_{c} = card area (area of indicator diagram, mm²)
   l_{c} = card length (length of indicator diagram, mm)
   k_{s} = spring constant (kPa/mm), which depends on the spring rate, the slave-piston area, and the magnification factor of the stylus; it is determined by the manufacturer of the indicator

The value of calculating mean effective pressure is that the working stroke may now be treated as a constant-pressure process. The indicated work may then be derived from Equation 3.6 using p_{m} as the constant pressure:

$$W = p_{\mathrm{m}} (V_2 - V_1)$$

If A = piston area (m²) and L = stroke (m), then the change in volume $V_2 - V_1 = AL$ and $W_{\mathrm{ind}} = p_{\mathrm{m}} AL$

Therefore: $$IP = p_{\mathrm{m}} L A n$$ indicated power (7.3)

where n = number of cycles per second.

For a multicylinder engine using p_m as a typical (or average) value for all cylinders, then

$$n = \text{number of cylinders} \times \frac{N}{60} \text{ (rpm) for two-stroke engines}$$

$$n = \text{number of cylinders} \times \frac{N}{120} \text{ (rpm) for four-stroke engines}$$

Example 7.2

A four-cylinder, two-stroke diesel engine running at 1000 rpm was tested with a mechanical indicator. The following results were obtained:

> average area of indicator diagram = 415 mm^2
> length of indicator diagram = 60 mm
> spring constant = 105 kPa/mm

The engine has a bore of 100 mm and a stroke of 120 mm. Calculate the indicated power.

Solution

$$p_m = \frac{A_c k_s}{l_c}$$

$$= \frac{415}{60} \times 105$$

$$= 726.25 \text{ kPa}$$

$$L = 0.120 \text{ m}$$

$$A = \frac{\pi d^2}{4} = \frac{\pi \times 0.1^2}{4} = 0.007\,85 \text{ m}^2$$

$$n = \frac{4 \times 1000}{60} = 66.67 \text{ cycles/s}$$

$$IP = p_m LAn$$

$$= 726.25 \times 0.120 \times 0.007\,85 \times 66.67$$

$$= \textbf{45.6 kW}$$

Self-test problem 7.4

A six-cylinder four-stroke engine with bore 120 mm and stroke 110 mm was tested with a mechanical indicator and the following results obtained:

> average area of the indicator diagram = 450 mm^2
> length of the indicator diagram = 50 mm
> spring constant = 90 kPa/mm
> engine speed = 800 rpm

Calculate the mean effective pressure and the indicated power.

7.6 FRICTION POWER

Indicated power is always greater than brake power; the difference between them is termed **friction power** (*FP*) because this power loss is due primarily to friction. Therefore:

$$\boxed{FP = IP - P}$$ friction power (7.4)

In a reciprocating engine, friction power is the power loss between the power developed at the face of the piston, and the power delivered by the crankshaft. It has a number of components:

- friction between moving parts such as the piston, connecting rod, crankshaft, camshaft and valve gear;
- air resistance due to the fact that the reciprocating or rotating components are surrounded by air (usually);
- power necessary to drive internal equipment such as the oil pump, fuel injection equipment, and electrical equipment;
- power absorbed by the cooling system in water pump and fan.

Because friction between lubricated surfaces depends on both the force and the rubbing speed, friction power increases with:

- speed of the engine at a given load (or power output);
- engine load (or power) at a given speed;
- compression ratio—compression-ignition engines have higher friction power than comparable spark-ignition engines;
- viscosity of the lubricating oil, which in turn depends on engine operating temperature.

7.7 MECHANICAL EFFICIENCY

The mechanical efficiency (η_m) is the ratio between brake power and indicated power.

$$\boxed{\eta_m = \frac{P}{IP}}$$ mechanical efficiency (7.5)

Mechanical efficiency decreases as friction power increases, and the higher the mechanical efficiency, the more efficient the engine is at converting the power developed by the working substance into useful power at the output shaft.

7.8 INDICATED THERMAL EFFICIENCY

Indicated thermal efficiency (η_{ind}) is the efficiency of the engine without frictional losses included, that is, the ratio between indicated power and heat-supply rate.

$$\boxed{\eta_{ind} = \frac{IP}{\dot{Q}_S}}$$ indicated thermal efficiency (7.6)

Indicated thermal efficiency is always greater than overall efficiency (brake thermal efficiency) because indicated power is always greater than brake power. Note that

$$\eta = \eta_{ind} \times \eta_m$$

since $\dfrac{P}{\dot{Q}_S} = \dfrac{IP}{\dot{Q}_S} \times \dfrac{P}{IP}$

Example 7.3

A heat engine develops 85 kW when the indicated power is 102 kW. The engine has a specific fuel consumption of 0.3 kg/kWh, and the fuel has an energy content of 46 MJ/kg. Determine:

(a) the friction power;

(b) the mechanical efficiency;

(c) the indicated thermal efficiency;

(d) the brake thermal efficiency (overall efficiency).

Solution

(a) Friction power:

$$FP = IP - P$$
$$= 102 - 85$$
$$= \mathbf{17\ kW}$$

(b) Mechanical efficiency:

$$\eta_m = \frac{P}{IP}$$
$$= \frac{85}{102}$$
$$= \mathbf{83.3\%}$$

(c) Heat-supply rate:

$$\dot{Q}_S = \frac{0.3 \times 85 \times 46}{60 \times 60}\ \text{MJ/s}$$
$$= 325.8\ \text{kJ/s}$$

Indicated thermal efficiency:

$$\eta_{ind} = \frac{IP}{\dot{Q}_S}$$
$$= \frac{102}{325.8}$$
$$= \mathbf{31.3\%}$$

(d) Brake thermal efficiency:

$$\eta = \eta_{ind} \times \eta_m$$
$$= 0.313 \times 0.833$$
$$= \mathbf{26.1\%}$$

Check:

$$\eta = \frac{P}{\dot{Q}_S}$$
$$= \frac{85}{325.8}$$
$$= \mathbf{26.1\%}$$

Self-test problem 7.5

The engine given in Self-test problem 7.4 has a mechanical efficiency of 85%. It consumes petrol with energy content 46.5 MJ/kg and relative density 0.78 at a rate of 11.3 L/h. Determine:

(a) the output power;

(b) the friction power;

(c) the efficiency (overall);

(d) the indicated thermal efficiency;

(e) the specific fuel consumption in kg/kWh.

7.9   VOLUMETRIC EFFICIENCY

The efficiency with which an engine induces air (or air–fuel mixture) is measured by the volumetric efficiency, which is the ratio of the actual volume induced divided by the theoretical maximum volume that can be induced:

$$\eta_V = \frac{\dot{V}_{actual}}{\dot{V}_{theoretical}}$$

Notes

1 For a positive-displacement engine, the maximum theoretical volume of air that may be induced per cycle is the swept volume of the engine. The theoretical volume-flow rate is the swept volume multiplied by the number of cycles per second.

2 When determining volumetric efficiency experimentally, an **air box** is usually necessary to smooth out airflow so that a steady reading of actual volume-flow rate can be taken.

3 Volumetric efficiency is improved by:

- reducing the engine speed;
- increasing the throttle opening (in spark-ignition engines);
- supercharging, that is, forcing the air into the engine using a blower or compressor; if the supercharger is powered by the energy in the exhaust gas (turbo charging), a significant gain in thermal efficiency is also achieved;
- reducing the air temperature prior to intake;
- attention to inlet conditions such as large (or more) inlet valves, smooth inlet passages, long valve-opening times, minimisation of sharp bends, large air filter, and so on.

 ## Self-test problem 7.6

A four-cylinder four-stroke spark-ignition engine has a bore of 80 mm and a stroke of 75 mm. When the engine speed is 4000 rpm, the air-flow rate to the engine is 32 L/s. Determine:

(a) the capacity of the engine;

(b) the volumetric efficiency.

7.10 *MORSE TEST*

The Morse test is an ingenious method of determining the indicated power of a multi-cylinder engine using only brake-power measurements. The test involves measuring the power output (brake power) of the engine and then disabling combustion in a cylinder by either shorting out a spark plug or overriding an injector. The engine speed will reduce; this speed is returned to its previous value by reducing the load on the engine. The power output is again measured and will be less than the initial power when all cylinders were functioning. The power difference is the indicated power of the disabled cylinder, as is now proved with the following example.

If all cylinders function in a four-cylinder engine:

$$IP \text{ (4 cyl.)} = P \text{ (4 cyl.)} + FP \text{ (4 cyl.)}$$

With one cylinder disabled, the friction power is assumed to be the same at the same speed. Therefore:

$$IP \text{ (3 cyl.)} = P \text{ (3. cyl.)} + FP \text{ (4 cyl.)}$$

Subtracting these equations:

$$IP \text{ (4 cyl.)} - IP \text{ (3 cyl.)} = P \text{ (4 cyl.)} - P \text{ (3 cyl.)}$$

Hence the difference in brake-power readings represents the indicated power of the disabled cylinder.

The *total* indicated power is obtained by repeating the test for each cylinder in turn, and adding the results.

The Morse test is a useful test but it somewhat underestimates the friction power (overestimates the indicated power) because the assumption of constant friction in the disabled cylinder is not really valid. When a cylinder is disabled, friction is less because piston forces are lower.

Example 7.4

A Morse test was performed on a four-cylinder spark-ignition engine connected to a hydraulic dynamometer. From the following results, determine the mechanical efficiency.

> engine speed = 3200 rpm
> dynamometer load arm = 350 mm
> dynamometer load with
>   all cylinders firing = 260 N
>   cylinder 1 not firing = 180 N
>   cylinder 2 not firing = 185 N
>   cylinder 3 not firing = 182 N
>   cylinder 4 not firing = 183 N

Solution

$$P = T\omega$$

$$\text{Brake power all cylinders} = 260 \times 0.35 \times \frac{\pi \times 3200}{30} \quad \text{W}$$

$$= 30.49 \text{ kW}$$

Average dynamometer reading with a disabled cylinder

$$= \frac{180 + 185 + 182 + 183}{4} = 182.5 \text{ N}$$

$$\text{Average } IP \text{ per cylinder} = 30.49 - \frac{182.5 \times 0.35 \times \pi \times 3200}{1000 \times 30} = 9.085 \text{ kW}$$

$$\therefore IP \text{ of engine} = 4 \times 9.085 = 36.34 \text{ kW}$$

$$\eta_\text{m} = \frac{P}{IP} = \frac{30.49}{36.34} = 0.839 = \mathbf{83.9\%}$$

7.11 *ENERGY BALANCE FOR A HEAT ENGINE*

The energy input to the engine must balance the energy outputs from the engine. For a heat engine that uses combustion of a fuel as a source of heat, the energy inputs and outputs are as shown in Figure 7.7.

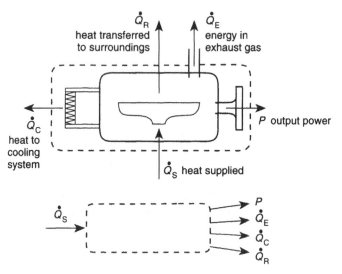

Fig. 7.7   *Energy balance for a heat engine*

Energy input—heat-supply rate

For an engine that utilises combustion of a fuel as the heat source, the rate of heat supply to the engine is given by Equation 7.1:

$$\dot{Q}_\text{S} = \dot{m}E$$

Energy outputs

The energy outputs from the engine are in the form of output power and energy dissipated (lost). The dissipated energy may be divided into three groups as shown in Figure 7.7. The three groups are now described.

Energy in the exhaust gas ($\dot{Q}_\text{E}$)   It is not practical to utilise all of the energy of combustion and a significant amount is lost in the exhaust gas.

One method of determining the energy rejected in the exhaust gas is to measure the mass-flow rate of exhaust gas $\dot{m}_\text{E}$ (equal also to the flow of air and fuel into the engine) and the difference between the temperature of the exhaust gas leaving the cylinder and the temperature of the air entering the engine, that is,

$$\dot{Q}_\text{E} = \dot{m}_\text{E} c_{p\text{E}} (T_2 - T_1)$$

where c_{pE} is the specific heat capacity of the exhaust gas at constant pressure (often assumed to be the same as for air).

An alternative method is to use a heat exchanger so that the heat of the exhaust gas is transferred to cooling water. From the energy balance of the heat exchanger, the heat rejected in the exhaust gas may be determined.

Note For an exact analysis, the energy content of any combustible emissions (hydrocarbons and carbon monoxide) should be taken into account when calculating the energy in the exhaust gas, as illustrated in Example 7.5.

Heat flow to the cooling system $(\dot{Q}_C)$ Due to the temperature limitations of metal components, cooling is necessary, and this is usually in the form of water cooling although some heat engines are air cooled.

If water cooling is used, the rate of heat flow through the cooling system may be determined by measuring the mass-flow rate of cooling water $\dot{m}_W$ and the temperature of the water entering and leaving the engine.
Therefore:

$$\dot{Q}_C = \dot{m}_W c_W (T_2 - T_1)$$

where c_W is the specific heat capacity of water (usually taken as 4.19 kJ/kgK).

Note   If air cooling is used, the same equation applies except that the mass-flow rate and the specific heat capacity of air should be used, rather than water.

Heat flow to the surroundings $(\dot{Q}_R)$ Not all the heat generated by friction, or occurring from heat transfer through the cylinder, transfers to the cooling water. Much heat also dissipates through the lubrication system and from the hot parts of the engine, and then to the atmosphere by conduction, convection and radiation. A small energy loss also occurs as a result of vibration, sound, and air resistance with the external moving parts.

Unless the engine is totally enclosed in an insulated container or room, the magnitude of the energy transferred from the engine to the surroundings is difficult to measure, and therefore is usually calculated by subtraction of the other energy outputs from the energy supplied by the fuel.

Example 7.5

Draw up an energy balance for a spark-ignition petrol engine, given the following test data:

> brake power = 35 kW
> fuel consumption = 0.18 kg/min
> energy content of fuel = 46.5 MJ/kg
> air consumption = 2.66 kg/min
> ambient air temperature = 20°C
> exhaust-gas temperature = 750°C
> cooling-water flow rate = 16.8 kg/min
> temperature rise of cooling water = 37°C

Assume the exhaust gas contains 0.5% hydrocarbons and carbon monoxide of average energy content 20 MJ/kg, and that c_p for the exhaust gas (c_{pE}) is the same as for air, namely 1.005 kJ/kgK.

Solution (working in kilowatts or kilojoules per second)

Energy in:

$$\text{Heat-supply rate } \dot{Q}_S = \dot{m}E$$

$$= \frac{0.18}{60} \times 46\,500$$

$$= \textbf{139.5 kW}$$

Energy out:

Brake power = **35 kW**

(a) Energy in exhaust gas.

    (i) Sensible heat:

$$\dot{Q}_E = \dot{m}_E c_{pE} (T_2 - T_1)$$

$$= \left(\frac{2.66 + 0.18}{60}\right) \times 1.005 \times (750 - 20)$$

$$= 34.73 \text{ kW}$$

    (ii) Combustion energy of emissions:

$$\dot{Q} = \dot{m}E$$

$$= \left(\frac{2.66 + 0.18}{60}\right) \times \frac{0.5}{100} \times 20\,000$$

$$= 4.73 \text{ kW}$$

$$\therefore \; \dot{Q}_E \text{ (total)} = 34.73 + 4.73$$

$$= \textbf{39.5 kW}$$

(b) Heat to cooling system:

$$\dot{Q}_C = \dot{m}_W c_W (T_2 - T_1)$$

$$= \frac{16.8}{60} \times 4.19 \times 37$$

$$= \textbf{43.4 kW}$$

(c) Heat to surroundings (by subtraction):

$$\dot{Q}_R = 139.5 - (35 + 39.5 + 43.4)$$

$$= \textbf{21.6 kW}$$

Energy balance

	kW	%
Supplied from fuel	139.5	100
Distributed:		
brake power	35.0	25.1
exhaust	39.5	28.3
cooling	43.4	31.1
remainder to surroundings	21.6	15.5

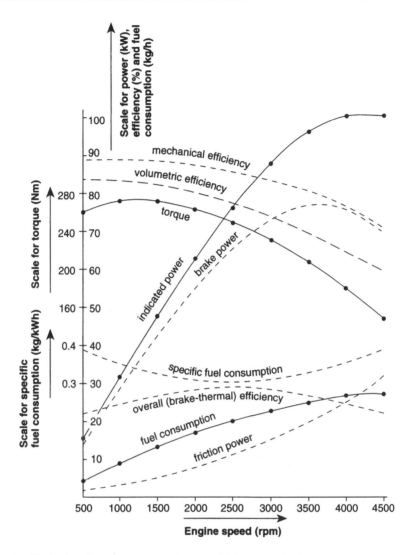

Fig. 7.8 *Typical performance curves for a variable-speed test*

7.12 *PERFORMANCE CURVES*

The performance data for a heat engine may be presented graphically. Each individual heat engine has its own characteristic performance curves and no two engines are exactly the same. However, similar heat-engine types produce similar performance curves, which are characteristic of the type. The shape of the curves may be altered by changes to the engine. For example, racing cars have specially modified engines so as to develop maximum power at high speed; however, their fuel economy (efficiency) is reduced.

Performance data may be obtained in two different ways:

- The engine speed is held constant and the load on the engine is varied. As the load is increased, the throttle is opened wider in order to maintain constant engine speed.
- The throttle opening is held constant at any given value and the speed of the engine is changed by varying the load on the engine.

Figure 7.8 (previous page) shows typical performance curves for a variable-speed test (the second type) at full throttle opening for a reciprocating spark-ignition internal-combustion engine.

Curves directly plotted from experimentally obtained data are shown solid, whereas curves obtained by calculation from the experimental data are shown dashed.

Torque

Typically, torque reaches a maximum value at fairly low engine speeds, after which the torque decreases as the engine speed increases. This is due to:

- volumetric losses (pumping and flow losses);
- friction losses that increase as the engine speed increases, and hence reduces the output torque.

Brake power

The brake-power curve is derived by taking the torque value at any speed and multiplying it by the angular speed. If the torque were constant, the brake power would be a straight line of positive slope, that is, the brake power would increase directly with engine speed. However, due to the nature of the torque curve, the brake-power curve reaches a maximum at fairly high engine speeds and then falls with increasing speed. For example, the engine tested in Figure 7.8 gives maximum power (at full throttle) at 3700 rpm.

Indicated power

The indicated-power curve has the same general shape as the brake-power curve, but the indicated-power curve is always greater than the brake-power curve at any given engine speed. Also, the indicated-power curve does not reach a peak as quickly as the brake power, and the maximum value occurs at a higher engine speed.

Friction power

The friction power is obtained by subtracting the brake power from the indicated power at any given speed. Friction power increases at an increasing rate with engine speed due to pumping and friction losses, which tend to increase approximately with the square of the engine speed.

Mechanical efficiency

The mechanical efficiency is obtained by dividing brake power by the indicated power at any given engine speed. As expected, mechanical efficiency decreases as the engine speed increases.

Fuel consumption

The fuel consumption increases with engine speed, but the rate of increase generally reduces as the engine speed increases.

Specific fuel consumption

The specific fuel consumption is derived by dividing the fuel consumption (kg/h) by the brake power at any given engine speed. Since fuel consumption and brake power both increase with engine speed, the specific fuel consumption does not show as marked a variation in value with engine speed. For the engine tested, it decreases from 0.38 kg/kWh at 500 rpm to a minimum value of slightly less than 0.3 kg/kWh at 2500 rpm, before increasing again to 0.38 kg/kWh at 4500 rpm. Therefore, the most economical operating speed is 2500 rpm.

Overall efficiency

The overall efficiency or brake thermal efficiency is derived by dividing the brake power (kW) by the heat-supply rate. The heat-supply rate is the fuel consumption (in kg/s) multiplied by the energy content of the fuel (kJ/kg). It reaches a maximum value when the specific fuel consumption is a minimum, confirming the fact that the most efficient operating speed for the engine tested is 2500 rpm. The maximum efficiency of slightly less than 30% is typical for the engine type.

Volumetric efficiency

Volumetric efficiency reduces as engine speed increases because of increasing fluid-flow losses. It has a maximum value of about 82% at idle speed, falling to about 60% at 4500 rpm.

Summary

There are many performance criteria applicable to heat engines of which the most important are torque, power and efficiency. The torque and power output can be measured in several ways, in which a resisting torque is applied to the engine by mechanical, hydraulic or electrical means. Power output is the torque multiplied by the angular speed. The efficiency of the engine is the ratio of the power output to the heat-supply rate. For an engine that burns a fuel, the heat-supply rate may be determined by measuring the flow rate of the fuel and the energy content of the fuel.

The indicated power of a heat engine is also an important performance measurement as it allows calculation of the friction power, mechanical efficiency and indicated thermal efficiency. Indicated power may be measured with a mechanical indicator (slow-speed engines) or an electrical or electronic indicator (high-speed engines). The indicated power of multicylinder engines may also be determined (fairly closely) with a Morse test.

The efficiency and power output of open-cycle engines are linked to the volumetric efficiency. Volumetric efficiency is determined by measuring the volume of air (or air–fuel mixture) drawn into the engine and dividing by the theoretical volume displaced.

The energy inputs and outputs of a heat engine must balance and an energy balance showing the input and output energies and how these are subdivided gives useful insight into the performance of heat engines.

Performance measurements are often presented graphically in the form of performance curves. These curves may be interpreted to determine the most economical operating speed, the maximum torque and power developed and the corresponding speed in each case.

Problems

7.1 A heat engine has a specific fuel consumption of 0.4 kg/kWh when the power output is 40 kW. Determine the overall efficiency, and fuel-flow rate to the engine (in L/min), if the fuel has energy content = 45.1 MJ/kg and relative density = 0.78.

    20%; 0.342 L/min

7.2 A large diesel engine used for propulsion of a ship has a maximum efficiency of 50% when producing 50 MW of power and when using heavy fuel oil with relative density 0.91 and energy content 42 MJ/kg. Determine:
(a) the specific fuel consumption in kg/MJ;
(b) the specific fuel consumption in kg/kWh;
(c) fuel consumption in L/h.

    (a) 0.0476 kg/MJ   (b) 0.171 kg/kWh   (c) 9420 L/h

7.3 With the aid of a *p–V* diagram, explain the meaning of *mean effective pressure*.

    A four-cylinder, four-stroke engine of bore 80 mm and stroke 90 mm has a power output of 50 kW at 4500 rpm. If the mechanical efficiency is 76%, calculate the mean effective pressure.

    970 kPa

7.4 A six-cylinder four-stroke engine has a bore of 110 mm and a stroke of 120 mm. When running at 640 rpm, an indicator card with an area of 420 mm^2 and a length of 60 mm is taken. If the spring constant is 120 kPa/mm, determine the mean effective pressure and the indicated power.

    840 kPa; 30.7 kW

7.5 An internal-combustion engine consumes 12 L of fuel in one hour while operating at a constant speed of 3000 rpm and turning an electric dynamometer. The dynamometer load reading is 200 N and the load arm length is 500 mm. The fuel has an energy content of 44 MJ/kg and a relative density of 0.78. Determine:
(a) the brake power;
(b) the specific fuel consumption;
(c) the overall (brake thermal) efficiency.

    (a) 31.4 kW   (b) 0.3 kg/kWh   (c) 27.5%

7.6 An eight-cylinder, two-stroke marine diesel engine has a bore of 750 mm and a stroke of 1120 mm and runs at 120 rpm. The indicated mean effective pressure is 620 kPa and the mechanical efficiency is 85%. Determine the indicated power and the brake power of the engine.

    4908 kW; 4172 kW

7.7 The marine engine given in Problem 7.6 is also produced in a double-acting version with the same bore, stroke and speed. The indicated mean effective pressure above the piston is 620 kPa, and below is 580 kPa. The piston-rod diameter is 250 mm and the mechanical efficiency is 82%. Determine the indicated power and the brake power of this version of the engine.

    8990 kW; 7371 kW

7.8 An indicator diagram is taken from a single-cylinder single-acting steam engine of bore 400 mm and stroke 500 mm, running at 250 rpm. The area of the diagram is 1380 mm^2, the length is 70 mm and the indicator-spring constant is 30 kPa/mm. Calculate the mean effective pressure and the indicated power of the engine.

The engine is coupled to a hydraulic dynamometer, which shows a brake load of 3.8 kN at a radius of 1.25 m. Determine the brake power and mechanical efficiency of the engine.

591.4 kPa; 154.8 kW; 124.35 kW; 80.3%

7.9 A four-cylinder two-stroke diesel engine is required to produce 100 kW at 2500 rpm. Design-indicated mean effective pressure is 1.15 MPa and mechanical efficiency is 80%. The engine is of a square design (that is, bore = stroke). Calculate the required cylinder diameter. Also, recalculate the necessary diameter if the engine were made four-stroke, all other conditions remaining the same.

94 mm; 118 mm

7.10 On test, a six-cylinder four-stroke engine at a speed of 4000 rpm showed a mean effective pressure of 1 MPa in each cylinder when the torque was 280 Nm. If the engine is of a square design with a bore of 100 mm, calculate the indicated power and the mechanical efficiency of the engine. If the engine consumed fuel, with energy content 43.5 MJ/kg and relative density 0.78, at a rate of 52 L/h, calculate also the specific fuel consumption and the overall efficiency of the engine.

157 kW; 74.7%; 0.346 kg/kWh; 23.9%

7.11 An eight-cylinder two-stroke compression-ignition engine has a bore of 100 mm and stroke 95 mm. When running at 2000 rpm, the quantity of air induced per second is 150 L. Determine:

(a) the capacity of the engine;

(b) the volumetric efficiency.

(a) 5.97 L   (b) 75.4%

7.12 A Morse test was performed on a six-cylinder spark-ignition engine running at 4000 rpm. From the following results determine brake power, indicated power, friction power and mechanical efficiency:

Dynamometer load:	all cylinders firing = 440 N
	cylinder 1 not firing = 345 N
	cylinder 2 not firing = 350 N
	cylinder 3 not firing = 360 N
	cylinder 4 not firing = 340 N
	cylinder 5 not firing = 350 N
	cylinder 6 not firing = 360 N
	dynamometer load arm length = 450 mm

82.9 kW, 100.8 kW, 17.9 kW, 82.2%

7.13 A rope of diameter 25 mm is wound around a flywheel of diameter 2.3 m. An engine turns the flywheel at 420 rpm against a mass of 60 kg attached to the lower end of the rope. The upper end of the rope is attached to a spring balance, which shows a reading of 12 N. Calculate the torque and power output of the engine.

670 Nm; 29.5 kW

7.14 An engine on test has a specific fuel consumption of 0.35 kg/kWh and uses fuel at the rate of 8.4 kg/h. The energy content of the fuel is 44 MJ/kg. Cooling-water flow to the engine is 12 kg/min, the temperature of the cooling water being 16°C at inlet and 53°C at outlet. The air–fuel ratio is 16 : 1, the ambient air temperature is 20°C and the exhaust-gas temperature is 680°C.

Draw up an energy balance for the engine on a one-minute basis. Assume c_p for the exhaust gas = 1 kJ/kgK and for cooling water = 4.19 kJ/kgK. Disregard the combustion-energy content of the exhaust emissions.

Q_S = 6.16 MJ; P = 1.44 MJ; Q_E = 1.57 MJ; Q_C = 1.86 MJ; Q_R = 1.29 MJ

7.15 The following table gives engine-test results for a variable-speed test at constant-throttle opening. The fuel had an energy content of 46.5 MJ/kg and an *RD* of 0.75. The dynamometer brake-arm length was 800 mm. Draw a set of performance curves for the engine (see Fig. 7.8), excluding volumetric efficiency. Show sample calculations for each curve drawn.

Speed (rpm)	1000	2000	3000	4000	5000
Brake load (N)	150	158	156	146	120
Fuel consumption (L/min)	0.123	0.194	0.233	0.281	0.374
Indicated power (kW)	15.7	31.7	47.2	62.7	70.8

7.16 A four-cylinder four-stroke petrol engine of 90 mm bore and 95 mm stroke gave the following results when tested at 3000 rpm:

        mean effective pressure:

                cylinder 1 = 1015 kPa

                cylinder 2 = 998 kPa

                cylinder 3 = 996 kPa

                cylinder 4 = 992 kPa

        nett brake load = 340 N

        brake arm = 460 mm

        fuel consumption = 20 L/h

Assuming the petrol has an energy content of 46.5 MJ/kg and a relative density of 0.74, determine:

(a) the indicated power;

(b) the brake power;

(c) the friction power;

(d) the mechanical efficiency;

(e) the specific fuel consumption;

(f) the indicated thermal efficiency;

(g) the brake thermal efficiency.

    (a) 60.45 kW   (b) 49.1 kW   (c) 11.3 kW   (d) 81.3%   (e) 0.3 kg/kWh   (f) 31.6%

    (g) 25.7%

7.17 A four-cylinder four-stroke internal-combustion engine was tested and the following data were obtained:

        bore = 85 mm

        stroke = 98 mm

        mean effective pressure = 720 kPa (average for each cylinder)

        nett brake load = 230 N

        brake arm = 450 mm

        fuel consumption = 9.4 L/h

        relative density of fuel = 0.76

        energy content of fuel = 43 MJ/kg

        engine speed = 2300 rpm

cooling-water flow = 24 L/min
temperature rise of cooling water = 14°C
water flow through exhaust-gas calorimeter = 32 L/min
temperature rise of water = 12°C

Assume the exhaust gas leaves the calorimeter at atmospheric temperature, and neglect the combustion-energy content of the exhaust emissions. Determine:
(a) the indicated power;
(b) the brake power;
(c) the mechanical efficiency;
(d) the specific fuel consumption;
(e) the indicated thermal efficiency;
(f) the brake thermal efficiency;
(g) the energy balance for the engine, and hence the energy to brake power, cooling, exhaust and atmosphere as a percentage of the input energy.
    (a) 30.7 kW   (b) 24.9 kW   (c) 81.2%   (d) 0.287 kg/kWh   (e) 36%   (f) 29.2%
    (g) 29.2%; 27.5%; 31.4%; 11.9%

7.18 A twin-cylinder four-stroke engine operates on natural gas of energy content at room conditions (15°C and 101.3 kPa) equal to 37 MJ/m^3. The bore of the engine is 95 mm and the stroke is 105 mm. At 4000 rpm, the engine develops 35 kW with a volumetric efficiency of 70%. The air–fuel ratio is 9.5 : 1 (by volume). Cooling-water flow is 25 L/min. The temperature rise of the cooling water is 21°C and the exhaust-gas temperature is 750°C. Assuming that c_p and c_v for the exhaust and inlet gases are the same as for air, draw up an energy balance for the engine and calculate the energy to brake power, cooling, exhaust and atmosphere as a percentage of the input energy. Assume there are no hydrocarbons in the exhaust gas.
    28.6%; 30%; 25.7%; 15.7%

7.19 **(a)** Draw typical performance curves for a spark-ignition internal-combustion engine at constant-throttle setting and variable speed. The curves do not have to be to scale but should show the general shape. All curves should be drawn on one sheet of graph paper. The following curves should be shown and marked:
(1) torque
(2) brake power
(3) indicated power
(4) friction power
(5) mechanical efficiency
(6) specific fuel consumption
(7) overall efficiency
(8) volumetric efficiency
(b) Explain how measured values may be obtained from an engine test that would enable each of the eight curves to be plotted and in each case outline the calculations necessary.
(c) Explain briefly the reason for the shape of each curve.

Fluid mechanics

Basic properties of fluids ≡

Introduction

Fluid mechanics is the study of fluids. It has several branches. The branch concerned with fluids at rest is known as fluid statics, and the branch concerned with fluids in motion is called fluid dynamics,. The study of pressurised air systems is known as pneumatics and the study of pressurised liquid systems is known as hydraulics.

This chapter is primarily concerned with the fundamental concepts and the basic properties of fluids; later chapters will be devoted to the study of the statics and dynamics of fluids in greater detail.

8.1 *TYPES OF FLUID*

A **fluid** is a liquid or a gas. At the molecular level, the distinction between fluids and solids is that in fluids the molecules have translational mobility whereas in solids they retain fixed spatial orientation. In solids, the molecules are able to vibrate or rotate only, but cannot change position relative to each other. The observable result of this is that fluids can flow whereas solids cannot.

A **liquid** is a fluid in which the molecules are free to move yet they still have *cohesiveness* because the molecules are close together and are attracted to each other. If a liquid is poured onto a flat surface, the liquid spreads out but remains in pools. A liquid poured into a container will settle with a free surface.

A **gas** is a fluid in which there is usually little or no cohesiveness between the molecules because they are widely spaced. If a gas is released from a vessel into the atmosphere, the molecules tend to disperse in all directions and do not cling together. A gas pumped into a vessel occupies the entire volume available with no free surface.

A **vapour** is a gas. However, the term vapour is usually used to refer to the gaseous state of substances that are usually liquid. For example, mercury is normally a liquid but there is mercury vapour present above the liquid surface in a mercury barometer. Similarly, water that evaporates is also known as water vapour. Solids can also be vaporised; for example, in a welding arc there is metal vapour.

A **slurry** is a liquid mixture containing suspended solid particles. The particles may be very fine as is the case when fine dusts are mixed with resins to make fillers or in the case of muddy water. The particles may also be quite coarse as is the case with blue-metal stones in concrete. Solid particles may also be suspended in gases as is the case with dust in air or the fine particles of carbon that appear as smoke in exhausts. Fine particles may remain in suspension for a long time but coarser particles tend to settle out under the influence of gravity.

Atomised liquids contain liquid particles suspended in a gas as fine droplets. For example, when liquid perfume, paint, fly spray or fuel is atomised, the liquid is converted to numerous fine droplets and mixed with air. The distinction between atomised liquids and vaporised liquids is that atomised liquids have not undergone a phase change and have not absorbed the latent heat necessary for the phase change. For example, steam can be superheated, dry or wet; superheated and dry steam are vapours whereas wet steam is vapour mixed with water droplets that have not yet undergone the phase change.

Foam (or froth) is produced when a liquid is agitated and large numbers of bubbles form. The bubbles are numerous pockets of gas within the liquid. Certain liquids are much more prone to foaming than others; for example, detergents foam very readily but mercury does not. Foaming adds to the difficulties of filling or pumping liquids but sometimes it serves a useful purpose (as with foam fire extinguishers or when clothes are being washed).

8.2 PROPERTIES OF A FLUID

Fluids have many characteristics or properties that are often classified as either physical or chemical properties. Some fluid properties important in engineering are now outlined.

The **solubility** of a substance in a liquid is the extent to which the substance will dissolve in the liquid to form a homogenous mixture. Unlike a slurry, separation cannot be effected by mechanical means (such as centrifuging). For example, when a cup of coffee is made with instant coffee, the coffee particles dissolve in the water, which is also the case if sugar is added. In carbonated beverages (such as soft drinks or soda water), the gas (carbon dioxide) is dissolved in water. Many liquids will also dissolve other liquids; for example, alcohol will readily dissolve in water (the basis of alcoholic drinks), whereas kerosene and oil are insoluble and will quickly separate out if mixed with water.

Whereas a solvent is any liquid that has dissolved another substance, the word solvent is also used to describe liquids that will dissolve solids readily. For example, many paints contain solvents that dissolve the colour pigments and other ingredients of the paint.

The amount of a substance that may be dissolved in the solvent (at given conditions) is known as the **concentration** and is limited to a maximum amount. Maximum concentration is known as the **saturated** condition. For example, if sugar is added to a cup of coffee until the water is saturated, then additional sugar will not dissolve but will remain in the solid state.

With solids or liquids, solubility depends primarily on the temperature of the solvent, and the higher the temperature the greater the solubility. For example, if salt is added to hot water until the saturation condition is reached, then, as the water cools, salt will come out of solution and form crystals.

With gases, solubility depends on both the temperature and the pressure of the solvent. The lower the temperature, the higher the solubility and the higher the pressure the greater the solubility. For example, if water is heated, dissolved air will be seen to come out of solution long before the water boils. Similarly, when a carbonated drink under pressure is opened, bubbles form as the carbon dioxide comes out of solution with the reduced pressure.

Solubility has significance in many engineering applications. For example, in a steam boiler, dissolved air coming out of the water is a nuisance because it reduces the heat-transfer capability and increases corrosion. Also, with propellers or pumps operating with liquids, gas bubbles in regions of low pressure may severely impede performance.

The **compressibility** of a fluid is the extent to which the fluid volume may be reduced by an increase in pressure. In liquids, the molecules are close together and, like a solid, cannot be made to close up much more even with the application of enormous pressures. Therefore, for most engineering applications in fluid mechanics, liquids may be regarded as *incompressible*, with little error.

However, the molecules of a gas are far apart and they can be brought closer together by compressing the gas and reducing the volume. Similarly a compressed gas can expand and increase in volume. There are many ways (or processes) by which expansion or compression of a gas can occur. These are dealt with in detail in Chapter 5.

The **surface tension** of a liquid is the cohesive force that occurs at the surface of the liquid. That is, the liquid surface appears to have a 'skin' or membrane over it. This can be verified experimentally by carefully placing a steel razor blade or a needle on a water surface, where it can be made to 'float'. However, if placed below the surface it will sink because steel is more dense than water. The apparent flotation is due to surface tension. Surface tension effects are also observable at the edges of a liquid in a container, where there is a 'meniscus'. With water the meniscus has an upward curve, whereas with mercury it has a downward curve.

Surface tension forces are small and can generally be neglected in fluid mechanics calculations. However, surface tension has some important effects; it is responsible for the spherical shape of bubbles and droplets of fluid and for 'capillary' action. When a small-bore tube (capillary) is inserted into a liquid that has an upward curving meniscus, the liquid will rise in the tube because there is an upward component of the surface tension force. This also causes 'wicking' where such a liquid will move of its own accord into a porous substance. This principle is used in the lubrication of bearings, enables ink to flow in ball-point pens and is the reason paper or cotton towels absorb liquids.

The **corrosiveness** of a fluid is the extent to which the fluid will enter into a chemical reaction with the materials it comes in contact with. For example, many acids react with many metals and are best kept in glass or plastic containers. Sea water is also corrosive and will react with many metals such as aluminium, copper and steel. Corrosiveness is very important in fluid mechanics because it dictates the materials suitable for containing or pumping fluids. For example, it would be poor engineering to specify aluminium or mild steel impellers for sea-water pumps because they would soon corrode and malfunction.

The **toxicity** of a fluid is the extent to which the fluid has harmful effects on humans (or other animals) when inhaled or digested. A toxic fluid is also known as a poisonous fluid. Toxicity depends primarily on concentration. For example, low concentrations of carbon monoxide in air cause no apparent health detriment, but as the concentration increases, there comes a point at which death can occur rapidly.

Toxicity is important to engineers because many fluid applications involve fluids that can be ingested by humans. For example, gases are often stored under pressure and a small leak of a toxic gas could increase the concentration of the gas to a dangerous level over a period of time. Also, fluids can dissolve substances that can accumulate over a period of time to toxic levels; for this reason lead pipes are no longer used for conveying water to be used for drinking or washing.

Other properties of fluids such as mass, volume, density and so on, are important physical properties that have already been treated in this book in Chapter 2. It is suggested that they now be revised under the following sections: 2.3 Mass, 2.4 Volume, 2.5 Density, 2.6 Relative density, 2.7 Specific volume, 2.8 Force, 2.9 Weight, 2.10 Pressure (absolute and gauge), 2.11 Temperature (absolute and Celsius). Two self-test problems now follow to help in this revision process.

 ## Self-test problem 8.1

A cylindrical vessel of internal diameter 560 mm and length 1200 mm contains 520 kg of liquid.

(a) Determine the density and relative density of the liquid.

(b) If the vessel held a gas of specific volume 0.35 m³/kg what mass of gas would there be?

 ## Self-test problem 8.2

A pressure gauge calibrator as shown in Figure 8.1 has a 30 mm diameter piston. The combined mass of the piston and carrier is 0.5 kg. Determine the correct pressure gauge reading when a mass of 1.5 kg is placed on the carrier.

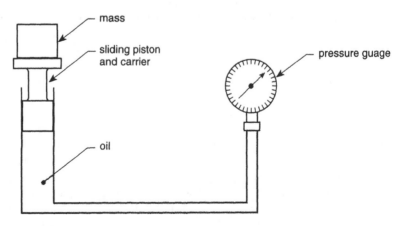

Fig. 8.1

8.3 *VISCOSITY*

The main distinction between fluids and solids is that fluids can flow whereas solids cannot. When a solid is moved over another solid, there is a friction force opposing the motion. Fluids also exhibit friction when the fluid flows through a pipe or from a hole in a tank. Some fluids flow more readily than others, and the resistance to flow is an important fluid property known as **viscosity**. For example, highly viscous liquids like honey or thick oil do not flow readily whereas low-viscosity liquids like water or alcohol flow far more readily. Gases also exhibit resistance to flow and have viscosity. As would be expected, the viscosity of liquids is far greater than the viscosity of gases.

In liquids, viscosity is due to cohesiveness, because the molecules tend to stick together and resist the shearing action necessary for flow to occur. Consequently, the viscosity of liquids decreases with temperature because cohesive forces reduce with temperature.

In gases, viscosity is due to momentum transfer, that is, collisions between faster and slower moving molecules, which have a component of the velocity in a direction perpendicular to the direction of flow. Because molecular velocity increases with temperature, the viscosity of gases (unlike liquids) increases with temperature.

Viscosity measurement

Viscosity can be measured in many different ways, but devices to measure viscosity (viscometers) can be grouped into one of three classes as outlined below. Theoretical calculation of the viscosity is impractical and viscometers need to be *calibrated* using fluids of known viscosity.

Types of viscometers

Rotational viscometers The fluid is enclosed as a thin film between two concentric cylinders. The outer cylinder is rotated at a given speed and the torque on the inner cylinder is measured. The greater the viscosity of the fluid, the higher the torque.

Falling-sphere viscometers A ball of standard size and weight is dropped a given distance through the fluid and the time measured. The higher the viscosity, the longer the time.

Flow viscometers A known quantity of fluid flows through a small orifice or capillary tube. The time taken is measured. The higher the viscosity, the longer the time.

Units of viscosity

There are two types of viscosity, namely kinematic viscosity and dynamic viscosity and each has different units.

Dynamic viscosity In the SI system, dynamic viscosity has units kg/ms, which is more often written as Pas (pascal seconds). In this book the Greek letter μ (pronounced mu) will be used for dynamic viscosity. The imperial unit of dynamic viscosity, the centipoise, is still in use. To convert from centipoise to Pas, multiply by 10^{-3}.

Note The SAE rating used with motor oils is the dynamic viscosity in centipoises at rated engine-oil temperature of 68°C. For example, SAE 30 oil has a viscosity of 30 centipoises at the rated temperature.

Kinematic viscosity In the SI system, kinematic viscosity has units m^2/s. In this book the Greek letter v (pronounced nu) will be used for kinematic viscosity. In imperial units, kinematic viscosity is measured in centistokes. To convert from centistokes to m²/s, multiply by 10⁻⁶. Kinematic viscosity can be calculated from dynamic viscosity by dividing by the density of the fluid as given by Equation 8.1 below:

$$v = \frac{\mu}{\rho}$$

kinematic viscosity (8.1)

Note   It is very important not to confuse kinematic and dynamic viscosity. Sometimes viscosity will be stated without qualifying whether it is dynamic or kinematic viscosity. In such cases, look at the units, which will define clearly which viscosity is being used.

 ## Self-test problem 8.3

An SAE 60 motor oil has a relative density of 0.92. Determine the dynamic and kinematic viscosity in SI units.

8.4   *SATURATION VAPOUR TEMPERATURE AND PRESSURE OF A LIQUID*

If water is placed in a container at atmospheric pressure and heat is supplied so as to raise the temperature, the water will boil at 100°C. This temperature is known as the *saturation vapour temperature*, because at this temperature the liquid water has absorbed all the energy it can without changing phase. That is to say, further energy input causes no change in temperature but rather a change in phase. If the pressure is reduced, and the experiment repeated, it will be found that the water boils at a lower temperature. If the pressure is increased, the water boils at a higher temperature (a fact used in pressurised-cooling systems and in pressure cookers). The relationship between saturation vapour temperature and pressure for water is shown graphically in Figure 8.2 (opposite).

A similarly *shaped* curve is obtained for most liquids but the *position* of the curve relative to the axes may be quite different from that of water. Liquids that boil more readily than water at lower temperatures (for the same pressure) are known as **volatile** liquids and include ether, alcohol and refrigerants. Indeed, the lower temperature of boiling of these liquids is the principle behind mechanical refrigerators. A typical curve for a volatile liquid (in this case a refrigerant, R12) is also shown in Figure 8.2.

8.5   *ENVIRONMENTAL IMPACT*

Fluids can escape into the atmosphere or into a waterway far more readily than solids and therefore their environmental impact is an important consideration for engineers. For example, when a car is hosed after being washed, detergents and oils in the wash water may flow into gutters and drains and eventually end up in a river or sea, where they will have deleterious environmental effects. Even non-toxic gases such as carbon dioxide or refrigerants have adverse ecological effects if released into the atmosphere in sufficient quantities (greenhouse effect).

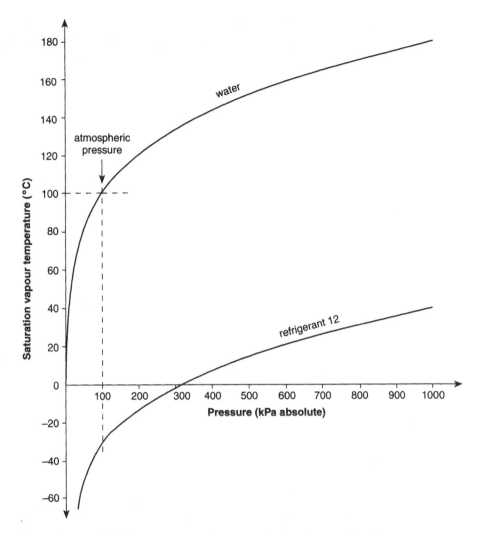

Fig. 8.2   *Relationship between saturation vapour temperature and pressure*

Disposal of liquids through sewage systems does not eliminate the environmental problems and is usually not allowed by governmental authorities. These systems are an essentially biological treatment of effluent and cannot eliminate heavy metals and other toxic substances. Therefore nowadays, most industrial plants are obliged to install and operate pollution-control equipment that reduces pollutant levels to an acceptable limit. Alternatively, liquids are often held in sludge tanks, then pumped out at periodic intervals and taken away to be dumped at a site where the ecological impact will be minimal.

Summary

A fluid is a liquid or a gas. Fluids can flow but exhibit resistance to flow in the same way that friction resists the movement of one solid over another. The resistance to flow of a fluid is known as its viscosity; high-viscosity liquids like honey do not flow readily whereas low-viscosity fluids like water flow readily. The viscosity of liquids is much higher than the viscosity of gases because of the cohesiveness of the molecules of a liquid. There are two types of viscosity, namely dynamic viscosity and kinematic viscosity and each has different units. Care should be taken to apply the correct viscosity and units when substituting in equations.

Often, fluids exist as a mixture of liquids, gases and solid particles. A liquid containing solid particles is known as a slurry whereas foams are a mixture of gases and liquids. Liquids may also be atomised and exist as numerous fine droplets.

Many solids, liquids or gases dissolve to some extent in a different type of liquid to form a homogeneous solution. The amount dissolved is known as the concentration (usually expressed as a percentage) and is limited to a certain maximum amount that occurs at the saturated condition. The concentration of a solid or liquid in another liquid increases with the temperature of the liquid. The concentration of a gas in a liquid decreases with the temperature of the liquid and increases with the pressure of the gas.

Liquids exhibit surface tension at the liquid surface. Surface tension forces are only small but are of importance in the operation of many devices. Surface tension causes capillary action, which enables liquids to 'wick' through porous substances and enables cotton or paper towels to absorb liquids.

Two important chemical properties of fluids are corrosiveness and toxicity. Corrosiveness needs to be taken into account when designing fluid-pumping or fluid-holding systems and toxicity is important whenever a fluid may be ingested or inhaled by people. Also care is needed to ensure that fluids do not escape into the atmosphere or waterways where they can have a deleterious environmental impact (particularly if they are toxic).

Many physical properties may be used to describe the state of fluids, such as mass, volume, density, and so on. Temperature and pressure are also important properties and care is needed to distinguish between absolute and Celsius temperatures and absolute and gauge pressures.

The temperature at which a liquid will boil (at a given pressure) is known as the saturation vapour temperature. A graph of saturation vapour temperature plotted against pressure shows that, as pressure rises, saturation vapour temperature also increases (but not linearly). Volatile liquids boil at lower temperatures than do other liquids.

 Problems

8.1 Describe briefly what is meant by the following:

(a) liquid

(b) gas

(c) slurry

(d) atomised liquid

(e) foam

8.2 Describe briefly what is meant by the following:
   (a) solubility
   (b) concentration
   (c) saturated solution

8.3 (a) Explain what is meant by the compressibility of a fluid; name the group of fluids that can be considered almost incompressible and explain why this is so.
   (b) Explain briefly what is meant by surface tension and describe an experiment to demonstrate the existence of surface tension.
   (c) Explain briefly what is meant by capilliary action and why it is of significance.

8.4 (a) Explain briefly the meaning of corrosiveness and toxicity of a fluid.
   (b) Explain briefly why engineers need to be concerned about environmental pollution when dealing with fluids and describe *two* types of environmental pollution that can occur with both gases and liquids.

8.5 (a) Explain briefly what is meant by viscosity of a fluid and how temperature affects the viscosity of both liquids and gases.
   (b) An SAE 30 oil has a relative density of 0.897 and a dynamic viscosity of 380 centipoise at 16°C. At this temperature determine:
   (i) the dynamic viscosity in SI units;
   (ii) the kinematic viscosity in SI units.
         (i) 0.38 Pas   (ii) 424×10^{-6} m²/s

8.6 Define the meaning of and state SI units for:
   (a) mass
   (b) volume
   (c) density
   (d) relative density
   (e) specific volume

8.7 (a) What is meant by the saturation vapour temperature of a liquid?
   (b) Draw a neat graph showing the saturation vapour temperature of water against absolute pressure, clearly marking the point of normal atmospheric pressure. Also show how the graph would look for a liquid more volatile than water.

8.8 A liquid half fills a cylindrical container with a diameter of 200 mm and a length of 1500 mm. If the weight of liquid is 223 N, determine the density and relative density of the liquid.
      965 kg/m³; 0.965

8.9 Gas of specific volume 0.6 m³/kg is in a cylindrical vessel of diameter 400 mm and length 1600 mm. Determine the mass of gas in the vessel.
   If a further 0.5 kg of the same gas is added to the vessel what will now be the specific volume and density?
      0.335 kg; 0.241 m³/kg; 4.15 kg/m³

8.10 Sea water of relative density 1.03 is at a depth of 2.5 m above a horizontal circular plate of diameter 400 mm. Determine the force on the plate and, hence, the pressure exerted by the sea water on it.
      3.17 kN; 25.3 kPa

8.11 A steel piston of diameter 50 mm and length 300 mm is placed in a vertical cylinder closed at the bottom end. Determine the pressure of the enclosed air when the piston comes to rest. Assume no friction or leakage loss and take the relative density of steel to be 7.8.

23 kPa (gauge)

8.12 Some of the trapped air is released from the piston- and cylinder-arrangement given in Problem 8.11 and the arrangement inverted. Determine the gauge and absolute pressure of the trapped air when the piston comes to rest once more.

–23 kPa (vacuum); 78.3 kPa

8.13 A steel tank of weight 4 kN has base dimensions of 1.2 m × 3.5 m and contains kerosene (*RD* 0.8) to a depth of 2.5 m. The tank is supported by four pads such that the load is equally distributed. Determine:

(a) the force on each pad;

(b) the fluid pressure on the base of the tank.

(a) 21.6 kN   (b) 19.6 kPa (gauge)

8.14 If 2100 L of kerosene is drawn out of the tank given in Problem 8.13, determine:

(a) the depth of kerosene now in the tank;

(b) the force on each supporting pad;

(c) the pressure on the tank base.

(a) 2 m   (b) 17.5 kN   (c) 15.7 kPa (gauge)

8.15 For the hydraulic jack illustrated in Figure P8.15, determine:

(a) the fluid pressure;

(b) the ram force *F*.

(a) 637 kPa (gauge)   (b) 5 kN

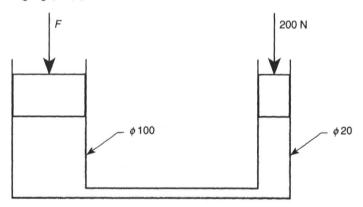

Fig. P 8.15

8.16 The pressure-gauge calibrator given in Self-test problem 8.2 has a mass of 2.5 kg placed on the carrier. The pressure gauge reading is 38 kPa. Determine the percentage error in the gauge reading.

8.73% too low

Components and their selection

Introduction

Fluid systems involve a number of components including pipes, fittings, valves, tanks and so on. The various common components found in fluid systems will now be discussed and their main features emphasised. The factors to be considered when selecting components are also listed.

9.1 PIPES, TUBES AND DUCTS

Pipes, tubes and ducts are used to transport fluid. There is no essential difference between a pipe and a tube, usage of one term over the other depends primarily on established practice. Ducts are usually made of thin-wall material such as sheet metal and are used for low-pressure gases (for example air-conditioning or ventilation ducts). Channels, culverts and drains are used to transfer a liquid (usually water) that is not under pressure but flows under the influence of gravity. In this case there is usually a free surface.

Section

The circular section is the most common because it has two outstanding advantages:

- There is maximum cross-sectional area to perimeter ratio. Therefore, for a given flow rate, the surface area of the pipe in contact with the fluid is a minimum. Frictional losses are therefore minimised.
- The circular section has maximum strength and requires a smaller wall thickness (for a given pressure) than any other section.

These two advantages mean that the circular section can convey the most amount of fluid with the minimum amount of pipe material, which in turn minimises cost and weight. However, in some cases other considerations dictate the use of a non-circular section. For example, when air-conditioning or ventilation ducts need to be installed in multi-floor buildings, the restricted space requires use of a rectangular or elliptical section in order to obtain the required flow rate.

Size

Pipes and tubes are manufactured in a large range of standard sizes, from very small capillary tubes to very large water and gas pipes. Standard manufactured sizes should be specified; often these are based on outside diameter and wall thickness, in which case the inside diameter is not a whole-number size and may not vary in uniform steps throughout the various sizes. In other cases, inside diameters are standardised.

Materials

Pipes and tubes are manufactured from a wide variety of materials, which can be classified into four main groups:

- ferrous metals: cast iron, plain carbon steels, alloy steels and stainless steels
- non-ferrous metals: copper, brass, aluminium, lead etc.
- non-metallic materials: glass, ceramics, concrete, plastic, rubber etc.
- composite materials: these are often used in order to couple desirable properties such as strength, corrosion resistance and finish; for example, steel pipes are often zinc galvanised or provided with an internal or external plastic liner and similarly, flexible pipes are often made of rubber or plastic with moulded-metal reinforcement or an external woven-metal-mesh cover, which increases the pressure rating

 ## *Self-test problem 9.1*

Determine the area/perimeter ratio of the following pipes:

(a) a circular pipe of diameter 100 mm;
(b) a square pipe of the same cross-sectional area as (a);
(c) a circular pipe of diameter 200 mm.

What conclusions can be drawn about the effect of pipe shape and size on their efficiency for conveying fluids?

9.2 PIPE FITTINGS

There are many standard fittings used with pipes and tubes. Fittings may be screwed, welded, brazed, bonded with an adhesive or soldered. (A range of copper fittings is available with the solder and flux already contained in the fitting, so that attachment requires only cleaning of the pipe and the application of heat.) The most common types are now briefly discussed.

Types of fittings

Sockets

Metal pipes are sometimes joined by welding or brazing the two lengths of pipe together, but this is not always practical or desirable. It is more common to use a socket, which is essentially a connector for two lengths of pipe of the same diameter. The joint is generally intended to be a permanent one and disassembly may be difficult (or impossible). Three different types of socket are illustrated in Figure 9.1.

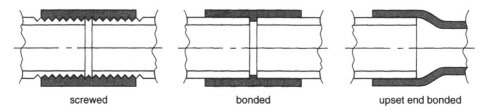

screwed   bonded   upset end bonded

Fig. 9.1   *Sockets*

Unions

Where two pipes are to be joined but the joint needs to be capable of easy disassembly, a union may be used. Unions have two halves, each of which is screwed or bonded to each length of pipe together with a tightening nut, which compresses the two halves together and seals them by means of a taper. A typical screwed union is illustrated in Figure 9.2.

tapered seal face ⌐   ⌐ tightening nut

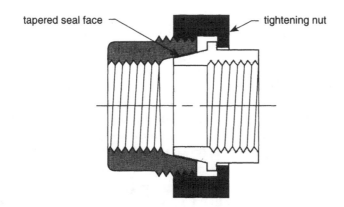

Fig. 9.2   *A typical union*

Unions are also used to connect pipes to equipment; for example, the connections to household water meters are made with screwed unions.

Flanges

Flanges perform the same function as unions but are used when high pressures are involved, particularly with steel pipes. Flanges are usually used with steam pipes and are also used to connect pipes to pressure vessels. Flanges may be welded or screwed to the pipes and mating flanges are bolted together (usually with four, six or eight bolts) using a gasket to ensure a leakproof seal between the faces.

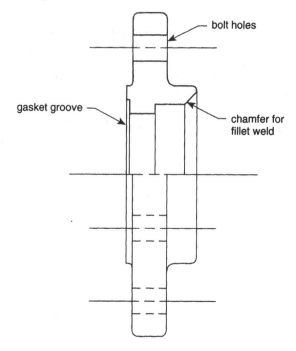

Fig. 9.3 *A pipe flange (welded)*

Compression fittings

Compression fittings are used with tubes of relatively small diameters. They are readily assembled and disassembled. Three main types of compression fittings are:

- A **flare fitting**, in which the end of the tube is swaged over using a flaring tool and is compressed between the two halves of the fitting to form a tight seal;
- A **ferrule fitting**, in which a small ring (also called an olive) is inserted over the tube and is compressed by the fitting to form a tight seal;
- an **O-ring fitting**, an O-ring (made of rubber or other pliant material) is inserted in a groove in the fitting and compressed against the tube to form a tight seal.

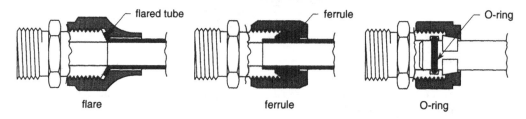

Fig. 9.4 *Compression fittings*

Transitions

A transition is used when two pipes of different diameters are to be joined. This is essentially the same fitting as a socket except that the diameters of each end are different.

To avoid flow loss, the transition should be gradually tapered, but sometimes for reasons of cost and size, this is not done.

Elbows and bends

Elbows and bends are used to obtain a change in direction. Standard types are illustrated in Figure 9.5. These are often available with various sweeps, for example close, medium or large sweep. The larger the sweep, the lower the flow loss but the greater the space needed and the higher the cost of the fitting.

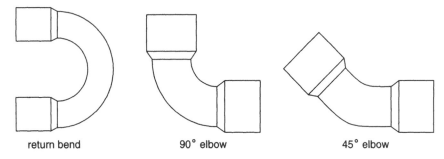

return bend   90° elbow   45° elbow

Fig. 9.5 *Elbows and bends (large sweep)*

Tee-pieces

Tee-pieces are used in order to connect a branch line into a pipe. The standard 90° tee is illustrated in Figure 9.6, but many variations are usually available. These include the 45° branch-line angle and the Y-connection in which both pipes diverge at 45° to the original direction. Usually there is a range of diametral changes available, with the branch line either the same diameter as the main line or with a reduced diameter.

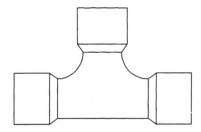

Fig. 9.6 *90° tee*

9.3 *VALVES*

Valves are used in fluid circuits for one or more of the following purposes:

- for on/off control (open/close, activate or isolate)
- to regulate (flow rate, fluid pressure or temperature)
- for directional control (divert the fluid from one path to another)
- for one-way control (prevent back flow or maintain prime)
- for safety (release pressure when set limits are exceeded)

Often a valve may serve a dual purpose but this is not always the case. For example, a valve that regulates flow will also usually provide satisfactory on/off control by throttling down to the fully closed position. However, a valve designed for on/off control only will not perform well if used for regulating.

Many different valves are available. Some valves are integral in the equipment; for example, motor-car engines have integral poppet or rotary valves, fuel pumps have integral one-way valves and carburettors have integral butterfly valves. The following discussion considers only the most common types of valves that are separate items of hardware installed in fluid systems.

Common types of valves

Gate valves

The gate valve, as illustrated in Figure 9.7, has a sliding gate, which moves up and down in guides. It is designed so that the gate wedges in the 'fully closed' position to effect a tight seal.

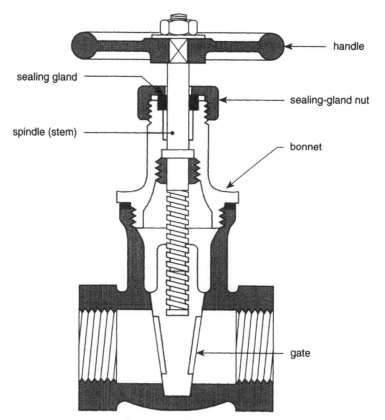

Fig. 9.7   *A gate valve*

Because of the straight-through configuration, gate valves cause a low flow loss in the 'fully open' position. Because of this, and the fact that they are relatively inexpensive, gate valves are widely used as isolating valves in fluid circuits. They are available in a wide range of sizes, for example bronze gate valves in sizes from 6 mm to 80 mm (nominal bore) and cast-iron gate valves in sizes from 40 mm to 600 mm. Gate valves are intended to operate only in the 'fully open' or 'fully closed' position and the practice of using gate valves for throttling should be avoided, since the gate is likely to 'chatter' in the partly open position, which causes damage to the valve.

An outstanding advantage of the gate valve is that it is suitable for high temperatures and pressures. The sealing gland can be made of a material with a high temperature and pressure rating. In any case, the sealing gland is located well away from the fluid and therefore remains relatively cool.

Globe valves

The globe valve, as illustrated in Figure 9.8, is widely used in fluid circuits where flow control is required (especially with manual operation).

The design of the valve is such that the fluid is forced through two sharp changes of direction, hence the flow loss is much higher than that of the same-sized gate valve in the 'fully open' position. However, the globe valve may be used at any intermediate position between 'fully closed' and 'fully open'. Standard sizes available are 8 mm to 175 mm (nominal bore).

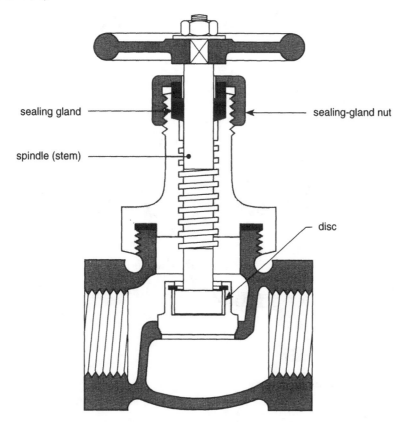

sealing gland

sealing-gland nut

spindle (stem)

disc

Fig. 9.8   *A globe valve*

Like the gate valve, the globe valve is suitable for high temperatures and pressures.

Needle valves

The globe valve does not give very fine control of the fluid flow (particularly at high pressures) because a small elevation of the disc causes a significant increase in the flow area. If fine control at low flow rates is required, a needle valve may be used. This valve is like a globe valve, with a tapered stem rather than a disc. Hence, as the valve is opened the increase in flow area is very gradual and thus fine control of the flow is achieved.

The head loss through a needle valve is very high and they are only used in relatively small sizes, 8 mm to 15 mm (nominal bore) being the standard size range available.

Diaphragm valves

None of the valves so far discussed is suitable for slurries containing abrasive particles because of the inevitable scoring and wear. If abrasive particles are present, the diaphragm valve may be used provided that the temperature does not exceed 175°C.

The diaphragm valve as shown in Figure 9.9 has a synthetic rubber diaphragm, which seals against a weir or seat. The resilience of the diaphragm allows for a tight seal on closure even when a solid particle is embedded in it. On reopening, the particle is passed down the line again and no permanent damage is caused either to the diaphragm or the seat.

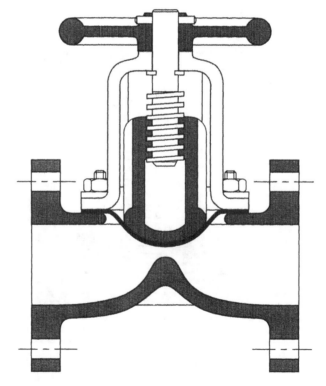

Fig. 9.9   *A diaphragm valve*

Ball valves

As illustrated in Figure 9.10, this valve has a spherical ball (usually made of stainless steel), in which there is an axial hole of diameter approximately equal to the inside pipe diameter. A stem fits into a slot in the top of the ball and is actuated by a handle, which requires only a 90° turn from fully open to fully closed. The ball is sealed by 'seats' (often made of nylon or teflon) and in some valves these may be tightened up to compensate for wear. Ball valves are widely used for on/off control and may also be used for throttling in some applications. However (due to the nature of the seat material), they are not suitable for high temperatures or high pressures. They are easily cleaned and hence are suitable for flow control in the food, drink and pharmaceutical industries.

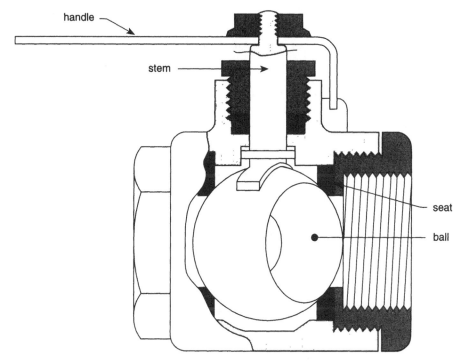

Fig. 9.10   *A ball valve (partly open position)*

Tapered-plug cocks

The tapered-plug cock is a simpler and less expensive version of the ball valve that uses a tapered plug instead of the ball. The tapered plug needs to be tightened enough to ensure a good seal but not too tight to hamper rotation. The plug may be made of a soft material such as brass or nylon or from a hard material such as ground glass. Tapered-plug cocks are suitable for low-pressure applications such as fuel-tank taps, gas taps and glass taps used in laboratory work. There are various versions of the tapered-plug cock. Some have O-ring seals at the top and bottom of the plug to increase the pressure rating and others have more than two ports and can be used for directional control.

Butterfly valves

The butterfly valve is similar in operation to the ball valve in that only a 90° turn is needed to effect full opening. A resilient seal may be used and hence, like the diaphragm valve, in such cases it is suitable for use when abrasive particles are present in the fluid. The valve is relatively cheap and lends itself to use with pneumatic, hydraulic or electric power actuators.

The main disadvantage of the valve is that it is only suitable for relatively low pressures and temperatures and a continuous force is necessary to hold it in the fully closed position (although in the off-centre design this force is partly provided by the upstream fluid pressure).

Check valves

Check valves or one-way valves are used to prevent backflow in a fluid circuit. They operate automatically and open when the fluid flows in one direction and close when flow is reversed. They are also used to maintain prime in a liquid pumping system.

There are many types of check valves, some are spring loaded and some use gravity to assist the closing action. The greater the closing force, the more rapid and positive the closing action but the higher the flow loss. A typical check valve is the swing check valve illustrated in Figure 9.11. This valve has a low flow loss because the disc swings out of the way when the valve is fully open. Other types include the ball-check valve and the lifting-disc type, both of which cause a higher flow loss than the swing-disc type because of the more tortuous fluid-flow path.

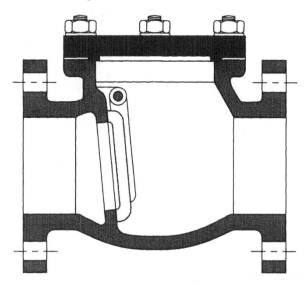

Fig. 9.11   *A swing check valve*

Foot valves

The foot valve is, as the name suggests, a valve placed at the foot or entrance of a pipe from a reservoir or tank. It is essentially a check valve designed to prevent the possible reversal of flow into the tank and also to maintain pump prime. Foot valves are manufactured in the same basic configurations as the check valve, namely swing-disc, lifting-ball or lifting-disc types. They are often fitted with a strainer to prevent particulate matter from entering the line.

Pressure-reducing valves

When there is fluid flow through any partly closed valve, flow is throttled and downstream pressure is lower than upstream pressure. When the downstream pressure needs to be regulated to a set value, pressure-reducing valves are used. These valves sense the downstream pressure and automatically adjust the valve opening to ensure that this pressure is held constant.

Safety valves

Safety valves are used where a safety risk exists with high-pressure fluids (particularly gases), as is the case with pressure vessels and boiler drums. The valves are kept closed by a spring load or weight load. When design pressure is exceeded, the opening force exerted by fluid pressure is greater than the closing force and the valve opens and relieves the excess pressure.

Directional control valves

Directional control valves are used in pneumatic or hydraulic circuits. They are usually designed for relatively small-bore tubing but for high pressures. They are used to control the motion of cylinders or motors and to perform fluid logic control. They may be operated manually or automatically in response to a mechanical signal, pressure signal or electrical signal. They usually have a sliding spool, which aligns with different ports as the position of the spool is changed.

Directional control valves are made in a huge variety of configurations, with different numbers of ports and flow characteristics. Some have a spring-loaded-return motion, others need to be actuated in both return and forward motions.

Further discussion of these valves is beyond the scope of this book and falls into the special area of fluid power and control.

 ## *Self-test problem 9.2*

List all the valves that would be suitable for the following applications:
(a) rapid on/off control;
(b) very accurate flow control;
(c) on/off and flow control of high-temperature and high-pressure fluids;
(d) to divert flow in hydraulic and pneumatic circuits;
(e) to prevent gas pressure from exceeding safe limits;
(f) to maintain a constant downstream pressure;
(g) to maintain prime for a centrifugal pump used with liquids;
(h) to control flow of a slurry containing small abrasive particles;
(i) to prevent backflow of a fluid.

9.4 *FILTERS AND STRAINERS*

Filters and strainers trap undesirable particles in fluids that could cause damage downstream. Filters trap fine particles whereas strainers trap coarse particles. Strainers are also used with pneumatic systems to remove water from the airstream. Filters are rated according to the smallest size of particle they will trap consistently. Very fine filters are able to remove smoke or airborne bacteria from air. Some filter and strainer elements may be cleaned by reversing the flow direction (backwashing), in other cases the element has to be removed and cleaned separately. Some elements (for example some made of paper) are not designed for cleaning and are simply throw-away items that are replaced when necessary.

It is good engineering practice to monitor the pressure drop across a filter, so that the filter element is replaced or cleaned at the appropriate time.

9.5   *STORAGE VESSELS AND TANKS*

Most fluid circuits include a storage tank or vessel. Common types are briefly discussed below.

Types of storage vessels and tanks

Weirs, dams, ponds and reservoirs

These are large storage vessels for water. Often they are natural ground depressions with a retaining wall to hold back the water.

Storage tanks

These are used for the storage of liquid at low pressure, such as water tanks and fuel tanks. This tank is often fitted with a vent to ensure that the tank vapour pressure does not exceed atmospheric and to enable the tank to be readily emptied or filled. Since the pressure is low, storage tanks may be constructed of relatively light material such as sheet metal or fibreglass.

Surge tanks

Surge tanks are often placed in the liquid line with the purpose of absorbing surges in liquid flow caused by fluctuations in flow rate or demand. The surge tank enables relatively steady flow to be maintained despite fluctuating-flow conditions (unsteady flow).

Header tanks

A header tank is similar to a surge tank but is included in the fluid circuit when it is necessary to maintain a constant head (or pressure) downstream of the tank despite fluctuations in head upstream of the tank.

Accumulators

Accumulators are used to perform the same function as surge tanks or header tanks except that they operate under high pressures and are used in hydraulic circuits.

Pressure vessels

Storage tanks for fluids under pressure are known as pressure vessels. Because of safety considerations they should be designed and constructed according to the appropriate codes and regulations. They require safety valves and inspection covers and the law requires that they are inspected at periodic intervals by a governmental authority.

9.6   *NOZZLES AND SPRAY HEADS*

Nozzles and spray heads are used in fluid systems where the fluid leaves a container or pipe under pressure and the fluid stream is directed at high velocity into a desired shape or pattern. Many different fluid shapes and patterns may be obtained but in all cases an increase in velocity is achieved by a reduction in the static pressure of the fluid. Sometimes, the nozzle is adjustable so that different patterns may be obtained with the same nozzle as is the case with watering-hose nozzles.

Converging nozzles concentrate the fluid into a coherent, high-velocity stream (also known as a jet). These are used in fire-hose nozzles and in gas, water and steam-turbine nozzles. Small-diameter water jets are used for a variety of purposes including water blasters for cleaning and water jets for cutting materials such as timber and fabrics.

Spray nozzles spread the fluid out into a fan or conical shape. This is used for many applications such as fuel injectors, spray guns and evaporative coolers. These nozzles usually have a centre section, which spreads out the fluid. Some also have small angled holes to help atomisation and to impart swirl to the fluid. Fuel-burner nozzles are often of this type.

Spray heads perform virtually the same function as spray nozzles except that they have multiple holes to spread out the fluid. Showers roses and watering cans usually use spray heads.

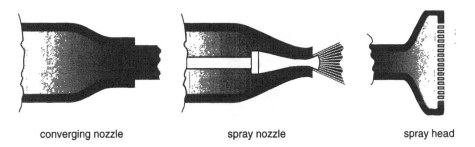

converging nozzle   spray nozzle   spray head

Fig. 9.12   *Nozzles and spray heads*

9.7 *GAUGES AND INSTRUMENTS*

There are many gauges and instruments available to measure fluid properties. It is good engineering practice to ensure that if instruments are installed as a fixture, they do not cause undue interference to fluid flow and that they can be removed for maintenance purposes without causing a major disruption.

Three most important properties of a fluid are pressure, temperature and level (volume). Instruments to measure these will now be discussed.

Pressure measurement

Fluid pressure can be measured in several ways as discussed below. If static pressure of a flowing fluid is being measured, any pressure tappings should be perpendicular to the direction of flow otherwise a false reading will be obtained.

Piezometers or manometers

Fluid pressure can be determined accurately from liquid height, which increases as pressure increases. From the measured height (and knowing fluid densities) the fluid pressure can be calculated. These devices and the calculations are given in Chapter 10 and will not be discussed further here.

Pressure gauges

The most common type of pressure gauge is the Bourdon-tube gauge illustrated in Figure 9.13 on page 202.

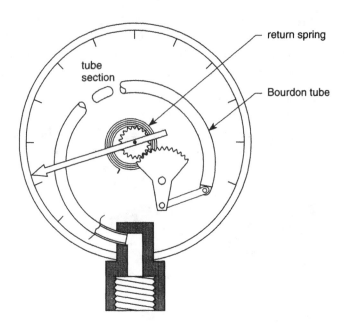

Fig. 9.13   *A Bourdon-tube pressure gauge*

The flattened tube (Bourdon tube) tends to straighten out as pressure increases and this movement is magnified by a sector and gear attached to a pointer. Pressure is read on a calibrated circular scale. This is a more direct, but less accurate, instrument for measuring fluid pressure than a piezometer or manometer.

Pressure transducers

Pressure transducers are small discs or plugs containing a material that compresses and changes electrical resistance as pressure increases. The change in electrical resistance can be converted to a voltage change, which can be calibrated as a pressure change. This can be displayed digitally or as a trace on a cathode ray oscilloscope. Pressure transducers are very compact and are manufactured in a wide variety of pressure ratings. They are usually used when rapid pressure changes occur, as these are difficult to measure by any other means.

 Self-test problem 9.3

What factors need to be considered when choosing a pressure gauge for any particular fluid-flow installation? Also, draw a neat sketch showing the correct method by which a pressure gauge should be fitted to a pipe.

Temperature measurement

Common methods of measuring fluid temperature are discussed below.

Thermometers

Thermometers are usually of the mercury-in-glass type. The expansion of the mercury relative to the glass is proportional to the temperature so that the glass stem may be calibrated in temperature units. Thermometers are accurate and inexpensive but they are fragile and sometimes difficult to read.

Fluid-filled gauges

A liquid-filled bulb is attached by a hollow stem to a pressure gauge. Because the vapour pressure of a liquid increases with temperature and is proportional to it, the pressure gauge can be calibrated to read temperature units.

Bimetal gauges

A bimetal gauge uses a coiled bimetal strip. Like a Bourdon tube, this strip tends to straighten as the temperature increases (because of the different expansion rates of the two metals). The movement is transmitted to a pointer and read on a dial calibrated in temperature units.

Thermocouples

A thermocouple is a junction of dissimilar metal wires connected to a voltmeter. As the temperature changes, the voltage changes. This voltage is proportional to the temperature and so the voltage reading can be calibrated in temperature units. The display can be analog or digital and it may also be presented as a continuous graphical readout.

Level measurement

When a liquid is resting in a tank with a free surface, it is often necessary to know how much liquid is in the tank, that is, the position of the liquid level. This may be done in several ways.

Sight glasses

The tank is fitted with a transparent vertical flat on one side. A conveniently sited scale enables liquid height to be read. Alternatively, a transparent vertical tube connected to the base of the tank may be used (in effect a piezometer). However, this will give a false reading of the liquid level if the tank is under pressure (non-vented).

Dipsticks

The tank is fitted with a dipstick, which can be withdrawn to read the level.

Floats

An internal float is connected through the tank to an external pointer and scale.

Pressure sensors

This is essentially a pressure gauge connected to the base of the tank but calibrated in units of height. This is not a suitable method if the pressure above the liquid in the tank varies (non-vented tank).

Electrical sensors

A float is attached to a wiper, which moves across a rheostat so that the electrical resistance varies with the position of the float. The change in resistance can be measured electrically and the instrument calibrated to read liquid height.

9.8   *FLOW MEASUREMENT*

The amount of fluid flowing (volume-flow rate) may be measured in two basically different ways, namely:

- from the volume displaced and the time taken;
- from the velocity of the fluid and the cross-sectional area.

Volume displaced method

This method is simple and accurate but does not give an immediate readout. The volume of fluid displaced may be measured in two ways:

- by diverting the flow through a measuring bulb or measuring cylinder;
- by installing a displacement meter (such as a gas meter, water meter or fuel meter), which gives a readout of the volume displaced.

The flow rate is obtained by dividing the displaced volume by the time taken. For example, if 6 L of fluid is displaced in one minute, then the volume-flow rate is 0.1 L/s.

Velocity of flow method

This method is adopted by many flow-measurement instruments that respond to the velocity of flow. The flow rate is then obtained by multiplying the velocity by the cross-sectional area. For example, if the velocity is 4 m/s and the cross-section area is 0.1 m^2 then the flow rate is 0.4 L/s. This calculation can be avoided and a direct readout obtained if the cross-sectional area does not change. In this case the scale may be calibrated in flow-rate units rather than velocity units.

Instruments using this method are described below.

Notches

If the fluid is a liquid and flows through a channel, a vee-notch as illustrated in Figure 9.14 is a simple and accurate means of measuring the flow rate. The height above the bottom of the notch is proportional to the flow rate and may be calibrated accordingly. Rectangular notches are also used in the same way, for example to determine the flow over a dam or weir by measuring the water height. Rectangular notches are less accurate than vee-notches at low flow rates.

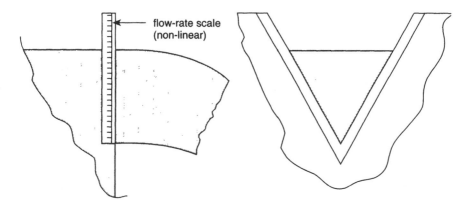

Fig. 9.14　*Vee notch*

Rotameters

A rotameter as illustrated in Figure 9.15 consists essentially of a vertical, tapered glass tube fitted with a float. The float may be spherical or specially shaped to rotate and maintain stability. As flow rate increases, the float rises and a scale at the side of the instrument is calibrated to give a flow-rate readout.

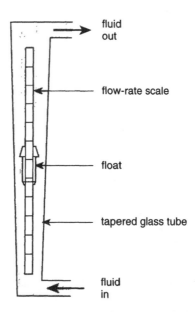

Fig. 9.15　*A rotameter*

Rotameters may be used with a variety of liquids or gases but the scale is accurate only for a particular fluid. However, different scales are usually supplied with each instrument to make it more versatile.

Venturi and orifice meters

These instruments consist essentially of a constriction in a tube, with pressure tappings at the constriction and upstream of it. In venturi meters, the constriction is tapered to minimise flow losses but in orifice meters, an orifice plate is inserted in the tube. This gives a high flow loss but enables the plate to be changed for different flow conditions.

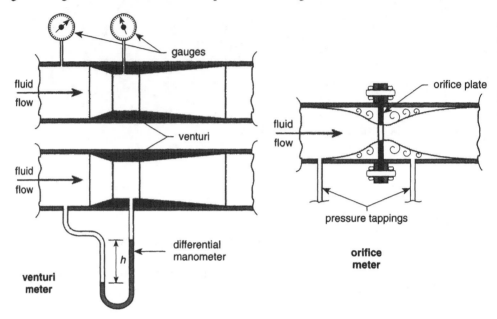

Fig. 9.16   *Venturi and orifice meters*

These instruments operate on the principle discovered by Daniel Bernoulli, namely that a constriction in the path of a flowing fluid causes an increase in velocity and a decrease in the static pressure. The pressure difference between the two pressure tappings can be measured and is proportional to the velocity (and flow rate), which may be calculated from the observed pressure difference.

Venturi and orifice meters are simple and robust and may be used with virtually any fluid. They are often fitted as permanent installations in pipelines. Their main disadvantage is that they do not give a direct readout of flow rate and calculation is necessary.

Pitot tubes

The Pitot tube illustrated in Figure 9.17 is a simple and versatile instrument for measuring the flow velocity. It is also used to measure the speed of aircraft or boats by measuring the flow velocity of the air or water past them. It has two tubes:

- an outer tube, with holes perpendicular to the direction of flow, which senses static pressure only;
- an inner tube, which faces into the direction of flow and senses the static pressure plus the pressure increase due to fluid striking the tube opening (dynamic pressure).

The dynamic pressure is greater than the static pressure and the pressure difference is proportional to the velocity. A differential manometer as illustrated in Figure 9.17 is often used with the instrument.

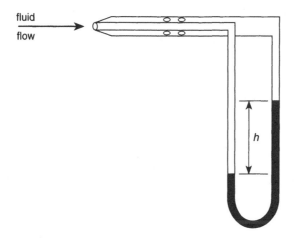

fluid
flow

h

Fig. 9.17 *A Pitot tube*

The difference in height of the indicating fluid in the manometer may be used to calculate the velocity and hence the flow rate by multiplying the velocity by the cross-sectional area. For greatest accuracy with large cross-sections, a traverse should be taken and a mean reading obtained.

Turbine meters

The turbine meter has a rotating blade (propeller, paddle or turbine). The rotating speed of the blade is measured (usually by electrical means) and this is proportional to the flow rate, which is displayed as an analog or digital readout.

Anemometers

As the name suggests, anemometers are used to measure the flow rate of air and are widely used for ventilation and air-conditioning measurements. They are portable and easily used across grills or ducts and require no installation. There are two types, namely fan anemometers and hot-wire anemometers.

The fan anemometer This instrument has a rotating fan in a cylindrical housing and operates in the same way as a turbine meter.

The hot-wire anemometer This instrument operates on the principle that the resistance of a wire changes with its temperature. A precise current is passed through a wire and, in still air, the wire attains a certain temperature. If the wire is now placed in an airstream, the equilibrium temperature attained depends on the flow rate of air over the wire. The resistance of the wire (and voltage drop across it) can be measured electrically and the output calibrated as an analog or digital display of flow rate or velocity.

As is the case with a Pitot tube, a traverse should be taken over large cross-sections for best accuracy.

9.9   *FLUID-POWER EQUIPMENT*

Fluid-power equipment can be divided into two classes:

* equipment that converts mechanical power (or energy) into fluid power (or energy) — pumps, compressors, fans, propellers and stirrers are in this class;
* equipment that converts fluid power (or energy) into mechanical power (or energy) — actuators in the form of cylinders, motors and turbines are in this class.

A detailed discussion of fluid-power equipment is beyond the scope of this text; however, a brief introduction follows.

Equipment that converts mechanical power into fluid power

Pumps

Pumps are used primarily with liquids and where a significant flow and pressure increase of the fluid is required. There are many types of pumps but the various types can be classed into one of two main groups: positive and non-positive-displacement types.

Positive-displacement pumps   These include piston pumps, gear pumps, gerotor pumps, lobe pumps, vane pumps and screw pumps. There are some ingenious designs such as swashplate pumps and bent-axis types where the volume-flow rate can be varied without changing the pump speed.

Rotodynamic or non-positive-displacement pumps   These are axial-flow, radial-flow or mixed-flow types, depending on the direction of the fluid leaving the pump.

Compressors

Compressors are used only for gases and are used when a pressure increase of the gas is the main requirement. Compressors are in the same two classes as pumps but the most common positive-displacement types are piston, vane and screw compressors whereas the most common rotodynamic types are centrifugal flow and axial flow.

Note that a pump effectively becomes a compressor if the delivery of the pump is connected to a pressure vessel in which the compressed gas accumulates and does not flow out.

Fans

Fans are designed to move large quantities of gas without achieving a significant pressure increase. A fan is virtually an axial-flow pump, used with gases.

Propellers

Propellers may be used with gases or liquids and work in the same way as axial-flow fans or pumps. However, their prime purpose is to obtain a large reaction force resulting from the momentum change of the fluid. This force is used to provide forward thrust to propel a boat or aeroplane.

Stirrers

Stirrers are used for stirring a fluid, that is, to mix or agitate the fluid as much as possible. Axial thrust is undesirable and one way of stirring fluids without developing thrust is to mount two propellers with opposite blade directions on the same shaft.

Equipment that converts fluid power into mechanical power

Actuators

Actuators can be divided into two main groups:

- continuous rotary motion—fluid motors and turbines;
- reciprocating linear motion—pneumatic and hydraulic cylinders.

Fluid motors and turbines There is no essential difference between a fluid motor and a turbine except that the term motor is usually used with positive-displacement types and the word turbine with the rotodynamic types. Most fluid pumps will also operate as motors if the flow direction is reversed, so that the fluid drives the motor or turbine in reverse.

Pneumatic and hydraulic cylinders The cylinder is fitted with a close-fitting piston driven by hydraulic or pneumatic pressure. A rod attached to the piston transmits the force to a mechanism or machine. The piston may be powered in both forward and return directions (two-way) or powered forward with a spring-loaded return (one-way).

9.10 *SELECTION OF FLUID COMPONENTS*

Many factors need to be taken into account when selecting components for a fluid system. Often there are conflicting advantages and disadvantages associated with each component and these need to be balanced so that the choice is optimal. The most important factors to be considered are listed below:

Volume-flow rate of fluid This affects the size of components when fluid flows through them. With pipes or valves, the greater the volume-flow rate, the larger the diameter needed. Also, when fluids are stored, the storage volume directly affects the size of the storage vessel.

Flow loss Flow losses represent lost energy and cause an increase in operating costs. Different components have different flow losses and, if possible, components that minimise flow loss should be selected.

Nature of the fluid In particular, whether the fluid is a liquid or a gas and whether it is corrosive or abrasive needs to be considered. Also, if the fluid is used for food, drink or pharmaceutical products, the material, surface finish and ease of cleaning of components are critical because of the need to maintain sanitary, contamination-free conditions.

Pressure of the fluid The fluid pressure directly affects material suitability and wall thicknesses.

Temperature of the fluid This also has a direct bearing on the materials that can be used. If large temperature variations occur it may be necessary to include expansion joints or loops.

The environment Components are often located outdoors and exposed to the weather, or located indoors, where there are some special environmental factors such as the presence of acidic vapours. Pipes and other components are sometimes buried underground or passed through a lake or ocean. Such environmental factors affect component suitability.

Vibration or movement   Fluid systems are often exposed to high vibration or movement, for example water pipes attached to a washing-machine bowl or hydraulic components on the wheel of a motor-vehicle. In such cases, components need to absorb these movements and still operate successfully over the life of the equipment.

Speed and response time   In some cases, components need to have a rapid response time. For example the shut-off time of a valve could be of prime importance in an emergency. The response time of instruments is often a key factor in deciding suitability.

Accuracy   This is a particularly important consideration in the choice of instruments used with fluids. For example, when choosing a pressure gauge, a gauge with a range of 0–100 kPa will be far more accurate than a gauge with a range of 0–1000 kPa if the pressure fluctuates between 20 and 80 kPa.

Convenience of installation and use   This affects the cost of installation, and the operating convenience. For example, a pressure gauge is a more convenient instrument to use than a manometer for measuring fluid pressure.

Reliability   Reliability is an important consideration in most fluid systems. It is of particular importance in applications such as aircraft pneumatic and hydraulic systems and motor-vehicle braking and steering systems.

Durability   Durability is a measure of the expected life of the component under operating conditions and is closely related to reliability.

Cost   Cost is almost always an important consideration. When comparing costs it is necessary to consider not only the initial cost of the components themselves but also their installation, maintenance and operating costs over the expected life of the installation.

Self-test problem 9.4

In an industrial plant, liquid is to be pumped from a storage tank through a system of pipes, fittings and instruments to an industrial process. If you were a design engineer on this project, what factors would influence your decision whether to use large or small pipes, fittings and instruments from the size ranges available?

Summary

A wide variety of components are used in fluid systems for the following purposes:

- To enable the fluid to flow through the system. This can be done in two ways:
  — open or free-surface flow—channel, culvert or drain
  — closed flow—pipe, tube or duct.

- To allow rigid pipes to be connected using pipe fittings such as a socket, union, flange, elbow, transition, tee or compression fitting.

- To control or regulate flow using a valve such as a gate valve, globe valve, needle valve, diaphragm valve, ball valve, tapered-plug cock, butterfly valve, check valve, foot valve, pressure-reducing valve or directional control valve.

- To reduce the risk involved with high-pressure gases using a safety valve.

- To store the fluid, which can be done in two ways;
  — in open vessels such as a dam, weir, pond, reservoir or open tank
  — in closed vessels such as a storage tank, pressure vessel, surge tank, header tank or accumulator.

- To obtain a high-velocity jet of fluid using a converging nozzle, or to spread the fluid out into a fan or conical stream using a spray nozzle or spray head.
- To measure fluid properties using instruments such as:
  — for pressure—piezometer, manometer, pressure gauge, or pressure transducer
  — for temperature—thermometer, fluid-filled gauge, bimetal gauge, or thermocouple
  — for level (or volume)—sight glass, dipstick, float, pressure sensor or electrical sensor.
- To measure fluid flow, which can be done in two ways:
  — by the volume displaced in a measured time, using a measuring bulb/cylinder, or a displacement meter
  — by the velocity and cross-sectional area, using a notch, rotameter, venturi/orifice meter, Pitot tube, turbine meter or anemometer.
- To convert mechanical power into fluid power using:
  — a positive-displacement pump such as a piston pump, gear pump, gerotor pump, lobe pump, vane pump or screw pump
  — a rotodynamic pump such as a centrifugal, axial-flow or mixed-flow pump
  — a compressor, which may be of the same type as a pump, but is used with gases to increase the pressure
  — a fan, which is usually an axial-flow machine designed to move large volumes of gases
  — a propeller, which is an axial-flow machine designed to obtain a large thrust.
- To convert fluid power to mechanical power using actuators, which can be:
  — a fluid motor or a turbine for continuous rotation
  — a pneumatic or hydraulic cylinder for reciprocating linear motion.

Selection of the appropriate component in any fluid system involves consideration of a number of factors and balancing their relative advantages and disadvantages to obtain an optimal solution over the life of the equipment.

 Problems

9.1 Describe *three* common methods for joining metal pipes and indicate two advantages and disadvantages of each method.

9.2 Briefly describe at least *one* application for the following types of valves: gate valve, globe valve, needle valve, diaphragm valve, ball valve, butterfly valve, tapered-plug cock, directional control valve, pressure-reducing valve, safety valve, check valve and foot valve.

9.3 With the aid of a neat sketch, describe the purpose and operating principle of the following valves: gate valve, globe valve and ball valve. Also describe their main advantages and disadvantages.

9.4 Outline the factors that should be considered when choosing the section, size and material of a pipe or tube to be used in a fluid-flow application.

9.5 With the aid of a neat sketch, describe the purpose and operating principle of the Bourdon-tube pressure gauge. Also describe the main advantages and disadvantages of this gauge compared with other methods of measuring pressure.

9.6 With the aid of a neat sketch, describe the purpose and operating principle of the following flow-measurement devices: vee-notch, rotameter, venturi meter, orifice meter and Pitot tube. Also describe their main advantages and disadvantages.

9.7 (a) List the *four* groups of materials out of which pipes and tubes are made and give at least *three* examples of the usage of each type in fluid-flow applications.
(b) Explain briefly why pipes and tubes are usually made with a circular section.
(c) Give a practical example of a non-circular duct or pipe used to transport a fluid and explain why this section is used.

9.8 Draw a neat cross-sectional view of the following fittings and state the purpose of the fitting: socket (screwed or glued), union, and flange.

9.9 Draw a neat cross-sectional view of the following fittings and state the purpose of the fitting: transition, 90° elbow or bend, and 90° tee.

9.10 Draw a neat cross-sectional view of the following compression fittings and explain how a seal is obtained with each fitting: flare, ferrule, and O-ring.

9.11 (a) Explain the purpose of a gate valve using an example.
(b) Explain whether or not a gate valve is suitable for throttling fluid flow.
(c) Explain whether or not a gate valve is suitable for high pressures and temperatures.

9.12 Describe *three* methods by which the temperature of a fluid flowing in a pipe may be measured. Indicate the relative advantages and disadvantages of each method.

9.13 Draw a neat sketch of a Pitot tube fitted with a differential manometer. Briefly describe the operation of the device, situations where it is commonly used and its relative advantages and disadvantages.

9.14 (a) What is the difference between a filter and a strainer?
(b) What *three* choices are available when a filter or strainer clogs?
(c) How can the clogging of a filter be monitored?

9.15 (a) What special considerations apply to the design, construction and use of gas-pressure vessels?
(b) Why are these considerations more important for high-pressure gases than liquids?
(c) Outline the purpose of the following vessels: surge tanks, header tanks, accumulators.
(d) Outline *four* different ways that the volume of liquid in a vessel can be measured.

9.16 (a) What are the two different methods used with instruments for measuring the flow rate of fluids?
(b) List *ten* different instruments that can be used to measure the flow rate of fluids.

9.17 (a) What are the *two* main classes of fluid-power equipment?
(b) Under each class briefly describe the various types of equipment.

9.18 (a) What is an actuator?
(b) What are the *two* main groups under which all actuators may be classed?
(c) What is the main difference between an actuator and a pump?
(d) Name an actuator that could be used as a pump and explain how this could be accomplished without changing the design of the actuator.

9.19 (a) What is the purpose of a nozzle or spray head?

(b) What changes in fluid properties occur when a fluid passes through a nozzle or spray head?

(c) Name *two* practical applications of converging nozzles, spray nozzles and spray heads.

9.20 List and briefly describe the various factors that should be taken into account when choosing fluid components for a particular application.

CHAPTER 10

Fluid statics

Objectives

On completion of this chapter you should be able to:

- state three basic principles of fluid statics that relate to fluid pressure at a point, at a wall and its transmission (Pascal's principle);
- recognise that the pressure at any horizontal level in a static liquid is the same and calculate that pressure;
- Explain, with the aid of a neat sketch, the use of piezometers and manometers (vertical, inclined and differential) and calculate pressures or pressure differences from the observed readings;
- calculate static-fluid-pressure forces and their location on surfaces (horizontal, vertical or inclined);
- state Archimedes principle and use it to calculate the magnitude and location of buoyancy forces on submerged, partially submerged or floating bodies.

Introduction

This chapter is concerned with fluid statics, that is, the principles applying to fluids at rest. Engineers are usually interested in static-fluid pressures and forces so this chapter concentrates on the relevant equations and principles that enable these pressures and forces to be calculated.

10.1 *BASIC PRINCIPLES OF FLUID STATICS*

Principle 1

The pressure at any point in a static fluid is the same in all directions.

Fluids flow when unbalanced pressures (and forces) are exerted on molecules of fluid. If the fluid is at rest, then, at any point, equal pressure (and force) acts in all directions.

Principle 2

The pressure (and force) at any point on the wall of a vessel containing a static fluid is perpendicular to the wall.

This principle results from the fact that when a fluid flows it exhibits viscosity or resistance to flow but when a fluid is at rest, there is no shear force (force component parallel to the wall). In a gas, the molecules impact the wall independently and cause reaction forces. The parallel components of these forces cancel each other out due to the huge number of impacts occurring in all directions, so the nett reaction force acts perpendicular to the wall. In a liquid, the molecules are free to slide over each other as if they were tiny balls, so the reaction force is also perpendicular to the walls.

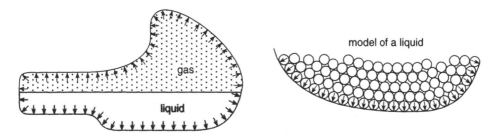

Fig. 10.1 *Pressure in a static fluid is perpendicular to the wall*

Principle 3 (Pascal's principle)

A pressure applied to an enclosed fluid or system of enclosed fluids at rest is transmitted without loss in all directions throughout the system.

Consider a system of enclosed fluids as shown in Figure 10.2.

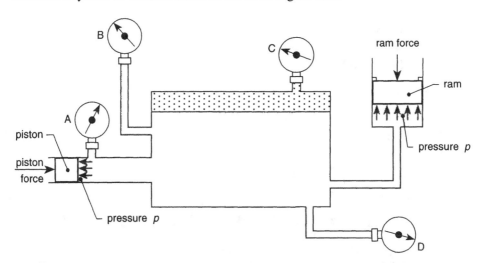

Fig. 10.2 *Pressure transmission through a fluid*

If the piston is held in position, each of the pressure gauges will register a certain pressure (which are not necessarily the same). If additional force is now applied to the piston causing an increase in pressure $p = F/A$, then the reading on each pressure gauge will increase by amount p regardless of where the gauge is located or how far away from the piston it is.

Force magnification or reduction

An important consequence of Pascal's principle is that a force can be magnified or reduced by means of fluid pressure. Indeed, this is the main reason why pneumatic and hydraulic systems are so widely used. Referring to Figure 10.2 it is seen that the piston has a small diameter so that a relatively small force can create a large pressure. However, the ram has a larger diameter and the same pressure acting on a larger area produces a larger force, that is, the force on the piston is magnified at the ram.

However, magnification of the force brings about a corresponding reduction in the distance moved. That is to say, the distance moved by the piston is greater than the distance moved by the ram.

Note   In Figure 10.2, gas has been shown above the liquid to demonstrate that Pascal's principle applies to both liquids and gases. However, in hydraulics, gas in the system is a nuisance because much of the applied pressure results in compression of the gas rather than movement of the ram. Hence, before use, hydraulic systems are bled to remove any gas.

Example 10.1

(a) The pressure gauges shown in Figure 10.2 read as follow: A 75 kPa, B 70 kPa, C 65 kPa, D 90 kPa. A force is now applied to the piston after which gauge A reads 90 kPa. What reading will be shown on each of the other gauges?

(b) If the piston has a diameter of 12 mm and the ram has a diameter of 80 mm, what force magnification will be obtained, that is, what is the ratio of the force at the ram to the force at the piston?

Solution

(a) Applied pressure = 90 – 75 = 15 kPa.

By Pascal's principle, each gauge will register 15 kPa higher, namely **B 85 kPa, C 80 kPa, D 105 kPa.**

(b) Using the subscript 1 for the piston and 2 for the ram:

$$p = \frac{F_1}{A_1} = \frac{F_2}{A_2}$$

Therefore $\dfrac{F_2}{F_1} = \dfrac{A_2}{A_1} = \dfrac{d_2^2}{d_1^2} = \dfrac{80^2}{12^2} = \mathbf{44.4}$

10.2 *PRESSURE VARIATION WITH DEPTH IN A LIQUID*

The static pressure at any horizontal position in a liquid due to the self-weight of liquid above increases in direct proportion to the depth. As shown in Figure 10.3 (case a), this implies a linear pressure distribution with depth, *regardless of the shape of the vessel.*

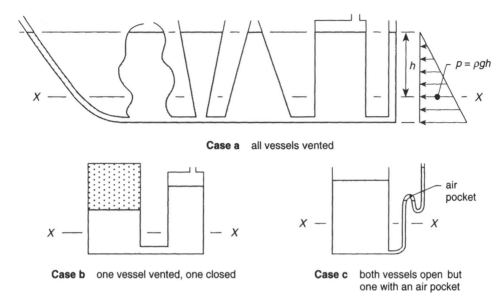

Case a   all vessels vented

Case b   one vessel vented, one closed

Case c   both vessels open but
one with an air pocket

Fig. 10.3   *Pressure distribution and surface level of a liquid in various connected vessels*

Two conclusions follow:

- *The pressure at any horizontal level in a static liquid in connected vessels is the same.* Referring to Figure 10.3, the pressure is the same at level X–X in each of the connected vessels in cases a, b and c. This conclusion follows because at the common connection point, the pressure must be the same and therefore at the same vertical height up from this point, the pressure must also be the same. This conclusion follows even if one of the vessels is closed as in case b and has a vapour pressure acting on the surface different to atmospheric pressure.
- *The surface in all open or vented connected vessels containing the same static liquid will be at the same horizontal level.* In Figure 10.3 (case a), the level is the same in all vessels because they are all open to the same atmospheric pressure. The surface levels are not the same in case b, because one vessel is closed and at a higher vapour pressure than atmospheric pressure. Also, the levels are not the same in case c because of the air-pocket in the right-hand limb.

The equation giving the static pressure at any depth due to the height of liquid below the free surface was quoted in Chapter 2 and is proved in Appendix 5. It is given by:

$$\boxed{p = \rho g h}$$ pressure increase due to self-weight of a liquid (2.8)

where p = pressure (Pa)
ρ = density of the liquid (kg/m^3)
g = gravitational constant (acceleration due to gravity) (N/kg or m/s^2)
h = depth below the free surface (m)

Notes

1 This equation applies to substances in which there are negligible density changes due to self-weight (incompressible). It applies to homogeneous liquids in any shape of

container but may also be used with homogeneous solids that have a uniform vertical cross-section throughout.

2 The pressure given by this equation is the *increase* in pressure due to self-weight. If the free surface is at atmospheric pressure and this is taken to be zero (gauge) then this equation will give the *gauge* pressure at depth h. If the free surface is at any pressure other than zero, the total pressure at depth h will be the sum of the pressures at the surface and the pressure increase due to self-weight at depth h. This is made clear in Example 10.2.

Example 10.2

For the oil tank shown in Figure 10.4, determine the gauge pressure p_2 when the pressure gauge p_1 shows (**a**) zero, (**b**) 25 kPa, (**c**) –10 kPa.

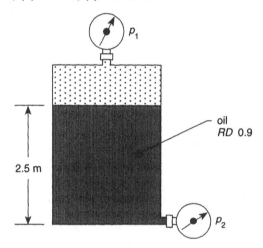

Fig. 10.4

Solution

$$p = \rho g h$$
$$= 0.9 \times 1000 \times 9.81 \times 2.5 \quad \text{Pa}$$
$$= 22.1 \text{ kPa}$$

(**a**)   If $p_1 = 0$
$$p_2 = \textbf{22.1 kPa}$$

(**b**)   If $p_1 = 25$ kPa
$$p_2 = 25 + 22.1 = \textbf{47.1 kPa}$$

(**c**)   If $p_1 = -10$ kPa
$$p_2 = -10 + 22.1 = \textbf{12.1 kPa}$$

 ### Self-test problem 10.1

Determine the atmospheric pressure when a mercury barometer shows a height of 752 mm. The relative density of mercury may be taken as 13.6. Assume a perfect vacuum (zero pressure) above the mercury. (Actually there is mercury vapour there but since the saturation vapour pressure of mercury at 20°C is 0.173 Pa, the error induced by neglecting this pressure is less than 0.0002%!)

10.3 *PIEZOMETERS AND MANOMETERS*

Piezometers and manometers provide a simple and accurate way of measuring relatively low static-fluid pressures. Piezometers are used only with liquids whereas manometers may be used with gases or liquids. If fluid flows in a pipe, piezometers and manometers may still be used to determine static-fluid pressures, provided that *the pressure tappings are perpendicular to the pipe.*

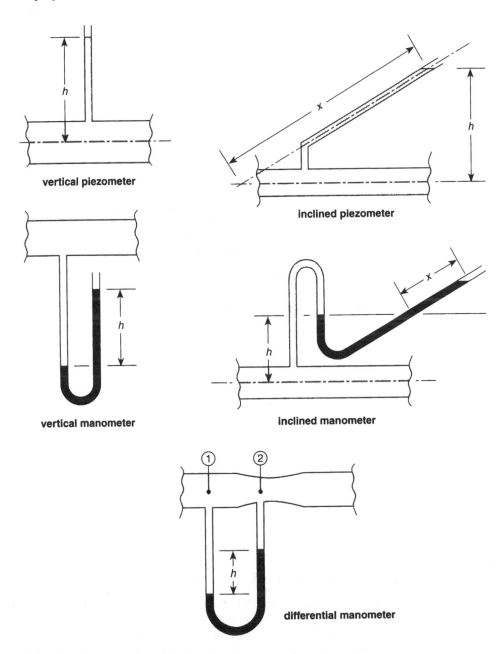

Fig. 10.5 *Piezometers and manometers*

Note   The static pressure in a pipe is always taken to be the pressure at the centre of the pipe unless otherwise stated.

Various types of piezometers and manometers are shown in Figure 10.5. Sometimes, for even greater accuracy at low pressures, one limb is inclined.

Vertical piezometers

Pressures may be calculated from a vertical piezometer reading using the formula $p = \rho g h$ as illustrated in Example 10.3.

Example 10.3

For the closed tank fitted with a piezometer as shown in Figure 10.6, determine the pressure at:
(a) the piezometer entry point;
(b) the base of the tank;
(c) the vapour space.

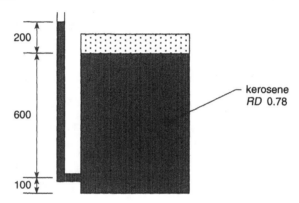

Fig. 10.6

Solution
(a) At piezometer entry, $h = 800$ mm $= 0.8$ m

$$\text{Now } p = \rho g h = 0.78 \times 10^3 \times 9.81 \times 0.8 \text{ Pa} = \textbf{6.12 kPa}$$

(b) At the base of the tank, $h = 900$ mm $= 0.9$ m

$$\text{and } p = 0.78 \times 10^3 \times 9.81 \times 0.9 = \textbf{6.89 kPa}$$

(c) The vapour-space pressure causes a rise in the piezometer of 200 mm, that is $h = 0.2$ m

$$\text{and } p = 0.78 \times 10^3 \times 9.81 \times 0.2 = \textbf{1.53 kPa}$$

Gas-to-liquid manometers

When a manometer is used to measure the pressure of a gas or vapour, the calculation is virtually the same as for a piezometer using height $h =$ difference in liquid height between one limb and the other. This is because in the limb containing gas, the density of the gas is negligible compared with the density of the liquid in the other limb. For practice in the calculation method, try Self-test problem 10.2, which follows.

Self-test problem 10.2

A mercury manometer is fitted to a glycerine storage tank as shown in Figure 10.7. Determine the vapour-space pressure and also the maximum tank pressure. Take the relative density of glycerine as 1.26 and mercury 13.6.

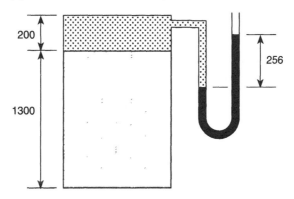

Fig. 10.7

Liquid–liquid manometers

If a manometer is used to measure the pressure of a liquid, there is liquid in both limbs and the density of both liquids needs to be taken into account in the calculation. Referring to Figure 10.8, between $X–X$ and $Y–Y$ there is the same liquid in both limbs and the pressure is the same at any horizontal level. Therefore, any horizontal level between $X–X$ and $Y–Y$ may be used as datum, but interface level $X–X$ is usually chosen.

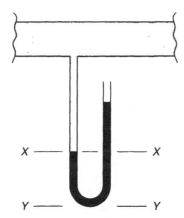

Fig. 10.8 *Any horizontal level between X–X and Y–Y can be used as datum*

The method of calculating the pressure is illustrated in Example 10.4.

Example 10.4

The pipe shown in Figure 10.8 is conveying water. Mercury is the indicating fluid in the manometer. The interface level $X–X$ is 400 mm below the centre of the pipe and the mercury level in the right-hand limb is 220 mm above the interface level. Determine the static pressure at the centre of the pipe.

Solution

Use X–X as datum and let p be the pressure in the centre of the pipe in kPa.

Left-hand limb:
$$\text{pressure at } X\text{–}X = p + \rho g h \text{ (water)}$$
$$= p \times 10^3 + 10^3 \times 9.81 \times 0.4 \text{ Pa}$$
$$= p + 3.92 \text{ kPa}$$

Right-hand limb:
$$\text{pressure at } X\text{–}X = \rho g h \text{ (mercury)}$$
$$= 13.6 \times 10^3 \times 9.81 \times 0.22 \text{ Pa}$$
$$= 29.35 \text{ kPa}$$

The pressure at X–X is the same in both limbs,
$$\therefore p + 3.92 = 29.35$$
$$\therefore p = 29.35 - 3.92$$
$$= \mathbf{25.43 \text{ kPa}}$$

Note The pressure calculated is a gauge pressure because atmospheric pressure above the mercury in the right-hand limb was taken to be zero.

Inclined piezometers or manometers

When a limb is inclined, the calculation follows the same method as for a vertical-limb instrument provided that vertical heights are used in the pressure calculation. For practice, attempt Self-test problem 10.3, which follows.

Self-test problem 10.3

A water manometer is used to measure the pressure of kerosene (relative density 0.78) flowing in a pipe as shown in Figure 10.9. Determine the static pressure at the centre of the pipe.

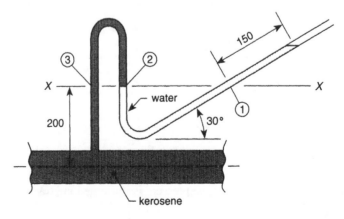

Fig. 10.9

Hint: Use X–X as datum and note that $p_1 = p_2$ and $p_2 = p_3$, therefore $p_1 = p_3$.

Differential manometers

A differential manometer (as illustrated in Figure 10.5) may be used to determine a pressure *difference* (hence the name). The calculation of pressure difference from the observed reading involves equating the pressures in the left-hand and right-hand limbs at the datum and is illustrated in Example 10.5, which follows.

Note   The distance of the manometer below the centreline of the pipe does not affect the calculated pressure difference because changes to this distance affect the pressure in both limbs equally and therefore has no effect on the *difference* in pressure.

Example 10.5

Water flows in a pipe to which a differential manometer containing mercury has been connected as illustrated in Figure 10.10. Determine the pressure difference between ① and ②.

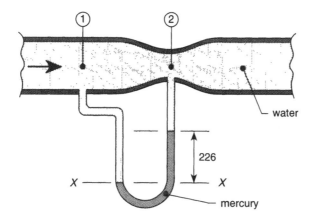

Fig. 10.10

Solution

Using *X–X* as datum, the pressure in the left-hand limb = the pressure in the right-hand limb. Remembering also that the distance of the manometer below the pipe does not matter, the equation is:

$$p_1 + \rho gh \text{ (water)} = p_2 + \rho gh \text{ (mercury)}$$
$$p_1 - p_2 = \rho gh \text{ (mercury)} - \rho gh \text{ (water)}$$
$$= 13.6 \times 10^3 \times 9.81 \times 0.226 - 10^3 \times 9.81 \times 0.226$$
$$= \mathbf{27.9 \ kPa}$$

10.4 STATIC-FLUID-PRESSURE FORCES ON SURFACES
Gas

When the fluid is a gas, the pressure does not vary over a surface exposed to that gas. The force in any direction due to the gas pressure may be calculated from the basic formula: $F = pA$. In this formula A is the *projected* area, that is, the area *perpendicular* to the direction of the force. The force is located at the centre (or centroid C) of the projected area. This is illustrated in Figure 10.11 on page 224.

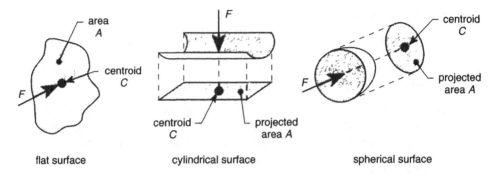

Fig. 10.11 *Force due to gas pressure*

Note This method may also be used with pressurised liquids when the pressure change due to self-weight is negligible. Example 10.6, which follows, illustrates this point.

Example 10.6

Oil pressure of 2.5 MPa is used to lubricate a split bearing of diameter 50 mm and length 75 mm. The bearing is split axially and the two halves are bolted together. Determine the force needed to hold the two halves together.

Solution

Projected area $A = 50 \times 75 = 3750$ mm$^2 = 3.75 \times 10^{-3}$ m^2

$$F = pA = 2.5 \times 10^6 \times 3.75 \times 10^{-3} = \textbf{9.375 kN}$$

Liquid on a horizontal surface

If the fluid is a liquid and the surface is a flat, horizontal one, the pressure is the same over the entire surface. The force is given by $F = pA$ and acts through the centre of area (centroid). For practice on the calculation method, attempt Self-test problem 10.4, which follows.

 ## Self-test problem 10.4

A circular plate of diameter 300 mm is screwed into the horizontal base of a vented tank containing an oil of relative density 0.9. If the depth of oil in the tank is 2.5 m, determine the force on the plate and its location.

Liquid on a vertical surface

In this case, the pressure on the surface is not constant because the pressure in a liquid varies with depth.

Magnitude of the force

The pressure distribution is linear and therefore the equations:

$$p = \rho g h \text{ and } F = pA$$

can be used provided that *p* is the *average* or mean pressure acting on the surface. That is, *p* is the pressure at the *centre of the area (centroid)*. This means that *h* is the vertical distance from the free surface to the *centre of the area* or *centroid*.

Location of the force

The force does not act at the centroid position but *below* it because pressure increases with depth. The position where the force acts is known as the **centre of pressure** and is below the centroid by a distance given by $y = k^2/h$, where *k* is the radius of gyration of the area. For a rectangle, $k = H/\sqrt{12}$ and for a circle $k = d/4$ where *H* is the height of the rectangle and *d* is the diameter of the circle.

Therefore, for a rectangular area:

$$y = \frac{H^2}{12h}$$ centre of pressure distance for a rectangle (10.1)

For a circular area:

$$y = \frac{d^2}{16h}$$ centre of pressure distance for a circle (10.2)

Note In the case of a vertical rectangle with the top edge at the surface, the centre of pressure is two-thirds of the way down the side (one-third the way up). This follows because of the triangular pressure distribution (see Figure 10.3).

Liquid on an inclined surface

Magnitude of the force

The vertical height to the centroid is still used to calculate the average pressure on the surface and hence the magnitude of the force by multiplying the average pressure by the sloping area. Using the subscript *i* to denote a dimension measured along the incline, the equations are:

$$p = \rho gh \text{ and } F = pA_i$$

Location of the force

Equations 10.1 and 10.2 may be used to locate the centre of pressure provided that *y* and *h* are measured *along the incline* and not vertically. Using the subscript i as before to denote dimensions measured along the incline, these equations become:

$$y_i = \frac{H_i^2}{12h_i} \quad \text{and} \quad y_i = \frac{d_i^2}{16h_i}$$

Note In the case of a sloping rectangle with top edge at the surface, the centre of pressure is two-thirds of the way down the sloping side (one-third the way up).

The method of calculating the force due to liquid pressure and its location is illustrated in Example 10.7.

Example 10.7

An 800 mm deep tank containing liquid of relative density 0.94 has one vertical side and one inclined side of length 1200 mm. Determine:

(a) the force on the vertical side per metre width of tank and its position relative to the base of the tank;

(b) the force on the inclined side per metre width of tank and its position relative to the base of the tank (along the incline).

Solution

Refer to Figure 10.12, where C is used to indicate the centroid and CP is used to indicate the centre of pressure.

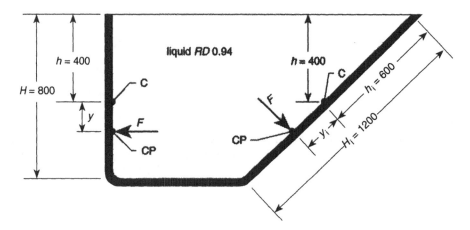

Fig. 10.12

(a) For the vertical side, $h = 400$ mm $= 0.4$ m

$$p = \rho g h = 0.94 \times 10^3 \times 9.81 \times 0.4 \text{ Pa} = 3.69 \text{ kPa}$$
$$F = pA = 3.69 \times 0.8 \times 1 = \textbf{2.95 kN}$$

Using Equation 10.1, with $H = 800$ mm $= 0.8$ m and $h = 0.4$ m

$$y = \frac{H^2}{12h} = \frac{0.8^2}{12 \times 0.4} = 0.133 \text{ m} = 133 \text{ mm}$$

That is, the force acts $400 - 133 = \textbf{267 mm above the base}$.

Note   Because the side is rectangular, with top edge at the surface, the force acts one-third of the distance up from the base, namely $800/3 = 267$ mm, which is the same result as obtained by use of Equation 10.1.

(b) For the sloping side the vertical distance to the centroid is the same and therefore the average pressure p is the same.

$$F = pA_i = 3.69 \times 1.2 \times 1 = \textbf{4.43 kN}$$

Using Equation 10.1 but measuring distances along the incline, $H_i = 1.2$ m and $h_i = 0.6$ m.

$$y_i = \frac{H_i^2}{12h_i} = \frac{1.2^2}{12 \times 0.6} = 0.2 \text{ m} = 200 \text{ mm}$$

That is, the force acts $600 - 200 = $ **400 mm from the base along the incline**.

Note Because the side is rectangular with top edge at the free surface, the force acts one-third of the distance up from the base, namely at distance $1200/3 = 400$ mm, which is the same result as that obtained with Equation 10.1.

 ### *Self-test problem 10.5*

(a) In the tank given in Example 10.7, a circular opening of diameter 300 mm is cut in a vertical side with the centreline of the hole at a distance of 200 mm from the base. Determine the force in an inspection plate bolted over the opening from the outside and its position relative to the base of the tank.

(b) Repeat part (a) but in this case the hole is cut in the sloping side of the tank at a centreline distance of 200 mm along the incline from the base.

10.5 *BUOYANCY FORCE*

A body immersed in a fluid experiences an upthrust force or buoyancy force. This force occurs because of the triangular pressure distribution in a fluid due to self-weight. The result of this is that the fluid pressure (and force) on the lower surfaces pushing up is greater than that on the upper surfaces pushing down. The side forces cancel each other (if they did not, the body would move of its own accord sideways!).

The magnitude of the buoyancy force is stated by **Archimedes' principle**, which states that: *A body wholly or partially immersed in a fluid experiences a buoyancy force equal to the weight of fluid it displaces.*

Notes

1 If the buoyancy force is greater than the weight of the body itself, the body will tend to move upward in the fluid until the forces balance. In the case of liquids, the body will float on the surface, displacing as much weight of liquid as its own weight.

2 Flotation occurs with homogeneous substances when the density of the substance is less than the density of the fluid. For example, oil or timber float on water because their density is less than the density of water (relative density less than 1).

3 Because the density of gases is low, buoyancy forces acting on bodies in gases are also low and are usually neglected in engineering calculations. For example, the buoyancy force on a person of average size due to air pressure is only about 1 N, which is the weight of a mass of about one-tenth of a kilogram. An exception to this are lighter-than-air balloons, which are able to rise because the buoyancy force is greater than their weight. In order for this to occur, the amount of air displaced must be large.

4 Archimedes' principle is proved in Appendix 13.

5 For a body that does not float at the surface, the difference between the actual weight of the body and the buoyancy force is known as the *apparent* weight because this is what the body appears to weigh when submerged.

6 The buoyancy force acts through the centre of mass of the displaced fluid and this point is known as the **centre of buoyancy**. The stability of a floating body depends on the vertical distance between the centre of buoyancy and the centre of mass of the body. For greatest stability, the centre of buoyancy should be high and the centre of mass should be low.

Example 10.8

(a) A steel block (*RD* 7.8) and dimensions 100 mm × 150 mm × 50 mm is lowered on a wire into a bath of oil (*RD* 0.9). What is the buoyancy force and the tension in the wire when the block is fully immersed?

(b) If the block is withdrawn from the oil bath and lowered into a mercury bath (*RD* 13.6) until it floats, what percentage of the block will be submerged?

Solution

(a) Displaced volume $V = 0.1 \times 0.15 \times 0.05 = 0.75 \times 10^{-3}$ m³

Buoyancy force = weight of oil displaced = $\rho_{oil} V g = 0.9 \times 10^3 \times 0.75 \times 10^{-3} \times 9.81$ = **6.62 N**
Weight of block = $\rho_{steel} V g = 7.8 \times 10^3 \times 0.75 \times 10^{-3} \times 9.81 = 57.4$ N
Wire tension = apparent weight = 57.4 − 6.62 = **50.8 N**

(b) When the block floats on the mercury surface, weight of block = weight of mercury displaced

$\therefore 57.4 = 13.6 \times 10^3 \times V \times 9.81$ where V = volume of mercury displaced
$\therefore V = 0.43 \times 10^{-3}$ m³

Therefore, percentage of block below surface = $\dfrac{0.43 \times 10^{-3}}{0.75 \times 10^{-3}} \times 100$
$$= \textbf{57.3\%}$$

Note This result could be obtained directly from the percentage ratio of the relative densities:

$$\frac{7.8}{13.6} = 57.3\%$$

 ## Self-test problem 10.6

A timber cube of side dimension 100 mm is placed in a cylindrical container of diameter 200 mm containing salt water (relative density 1.03). The level in the container rises 24 mm when the block floats. Determine:

(a) the mass of the block and hence the density of the timber;
(b) the downward force necessary in order to fully submerge the block.

Summary

The three basic principles of fluid statics are: pressure at any point in a fluid is the same in all directions, pressure at the wall of a vessel containing fluid is perpendicular to the wall at any point, and pressure is transmitted throughout a static fluid without loss (Pascal's principle). In a vessel or system of open, interconnected vessels containing a homogeneous liquid, the pressure at any horizontal level in each vessel is the same regardless as to the shape or size of the vessel. The corollary is that the liquid will find the same level in all vessels. The pressure at any depth in a liquid due to self-weight is given by the formula $p = \rho gh$ and therefore the pressure distribution is linear.

Piezometers and manometers may be used to accurately determine static-fluid pressures. The equation $p = \rho gh$ is used to calculate the fluid pressure from the observed height. With manometers, the method involves balancing the pressures in the left-hand and right-hand limbs at any convenient level, usually taken as the interface level.

If the fluid pressure acting on a surface is constant (or approximately constant), the force on the surface due to this pressure is calculated by multiplying the pressure by the *projected* area in the direction of the force. The force acts through the centroid (centre of area) of the projected area. When liquids are stored in tanks, the pressure change due to self-weight is usually significant and should not be neglected. In such cases, the magnitude of the fluid-pressure force on any plane area may be calculated by multiplying the *average* pressure by the area. If the area is vertical or inclined, the force no longer acts at the centroid but at the **centre of pressure**. This is always *below* the centroid by an amount that may be calculated for rectangular or circular areas using Equations 10.1 and 10.2 given in the text. Remember that when the area is inclined, the calculation to locate the centre of pressure involves using inclined distances and not vertical ones. However, vertical distances are still used to calculate the average pressure on the inclined surface. If the area is of a shape other than rectangular or circular, it is necessary to calculate the radius of gyration and use the formula $y = k^2/h$ to locate the centre of pressure.

Archimedes' principle states that the buoyancy force experienced by a body in a fluid is equal to the weight of fluid the body displaces. If the weight of the body is greater than the buoyancy force, the body will sink. However, it will appear to be lighter, that is, have an apparent weight less than the actual weight. If the buoyancy force is greater than the weight of the body it will rise and, if the fluid is a liquid, will float at a level such that the displaced weight of fluid equals the weight of the body, that is, the buoyancy force equals the weight. Buoyancy force acts through the centre of buoyancy, which is the centre of mass of the displaced fluid. For maximum stability of a floating body, the centre of buoyancy should be as high as possible and the centre of mass of the body as low as possible.

Problems

Note In all cases assume that pressures are gauge pressures (unless otherwise stated) and that the *RD* of mercury is 13.6.

10.1 (a) State the *three* principles of fluid statics that relate to fluid pressure: at a point, at a wall and its transmission.

(b) Draw a neat sketch showing how the pressure varies with depth in a liquid.

(c) What effect does the shape of the vessel have on the pressure distribution?

(d) Several vented vessels of different shape are interconnected by tubes at their bases. If these vessels all contain the same static liquid, what *two* conclusions can be drawn and under what circumstances are these conclusions invalid?

10.2 In a hydraulic system, a reciprocating piston pump has a diameter of 13 mm, a stroke of 20 mm and a piston force of 150 N. Pressurised oil is supplied to a ram of diameter 100 mm. Determine:

(a) the hydraulic pressure;

(b) the force at the ram;

(c) the number of cycles of the piston needed to move the ram 500 mm.

   (a) 1.13 MPa   (b) 8.88 kN   (c) 1479

10.3 (a) In a liquid of relative density 1.2, the surface is at a pressure of 30 kPa (gauge). What will the gauge and absolute pressure be at a depth of 3 m?

(b) A spherical vessel of inside diameter 800 mm contains gas at a pressure of 1.2 MPa (abs). It is made in two halves, which are bolted together. What force is induced in the bolts by the gas pressure?

   (a) 65.3 kPa (g) 166.6 kPa (abs)   (b) 552 kN

10.4 (a) Draw a neat sketch of a mercury barometer and explain how it works.

(b) How does the saturation vapour pressure of mercury affect the accuracy?

(c) If atmospheric pressure is 99.5 kPa, what height will be shown by the barometer?

   (c) 745.8 mm

10.5 (a) State Archimedes' principle.

(b) Explain why buoyancy forces occur and state where they act.

(c) When a body made of homogeneous material floats in a liquid, what conclusions can be drawn?

(d) Explain why a steel ship is able to float.

(e) What is the condition for maximum stability of a floating body?

10.6 (a) A piezometer is used to measure the pressure of water in a pipe. The water level in the piezometer is 200 mm above the centreline of the pipe. Determine the pressure at the centre of the pipe.

(b) A piezometer is used to measure the pressure of a fluid (*RD* 1.2) in a pipe of outside diameter 200 mm. The fluid level in the piezometer is 256 mm above the outside of the pipe. Determine the pressure at the centre of the pipe.

   (a) 1.96 kPa   (b) 4.19 kPa

10.7 An inclined piezometer slopes at an angle of 30° to the centreline of a horizontal pipe carrying water. The piezometer reading is 320 mm (measured from the centreline of the pipe along the incline). What is the pressure at the centre of the pipe?

   1.57 kPa

10.8 A pressure gauge attached to a vessel reads 50 kPa on a day when the mercury barometer reading is 774 mm. Determine the absolute pressure in the vessel.

   153.3 kPa

10.9 (a) A spherical buoy of diameter 500 mm floats in sea water half submerged. Determine the weight of the buoy. The *RD* of sea water = 1.03.

(b) Determine the force that would be necessary to just raise a steel block of dimensions 2 m × 1 m × 1.5 m resting at the bottom of the sea. The *RD* of sea water = 1.03, the *RD* of steel = 7.8.

(a) 331 N   (b) 199 kN

10.10 In an experiment, a cubic block of aluminium of side 200 mm is weighed. It is then placed in an evacuated chamber and weighed again. If the weights were 211.91 N and 212.00 N respectively, determine the relative density of aluminium and the density of air as indicated by this experiment.

2.7; 1.15 kg/m³

10.11 At a certain distance below the surface of a liquid the pressure is 100 kPa. Eight metres further down the pressure is 200 kPa. Determine the relative density of the liquid.

1.274

10.12 A pipe of internal bore 400 mm contains water at rest. The pressure at the centre of the pipe is 2.5 kPa. Determine the pressure at the top and bottom of the pipe.

0.54 kPa; 4.46 kPa

10.13 A pipe containing water at rest slopes upward at a slope of 1 in 100 (sine). The pressure at a certain point in the pipe is 80 kPa. Determine the pressure 1 km up the slope and 1 km down the slope from this point.

−18.1 kPa; 178 kPa.

10.14 A balloon, when inflated, is a sphere of diameter 8 m. The deflated balloon and attachments have a mass of 100 kg. Calculate the payload (mass-carrying capacity) of the balloon when inflated with hydrogen. Take the specific volume of atmospheric air to be 0.8 m³/kg and that of hydrogen to be 12 m³/kg.

213 kg

10.15 A mercury manometer is used to measure the pressure of oil (*RD* 0.9) in a pipe. The interface level is 300 mm below the centreline of the pipe and the level in the open limb is 125 mm above interface level. Determine the oil pressure in the pipe.

14 kPa

10.16 A mercury manometer is used to measure the pressure of water flowing in a pipe. The interface level is 245 mm below the centreline of the pipe. Determine the pressure in the pipe when the mercury level in the other limb is:

(a) 206 mm below the centreline;

(b) 294 mm below the centreline.

(a) 2.8 kPa   (b) −8.94 kPa

10.17 An inclined water manometer is used to measure the pressure of gas in a main. The open limb is inclined at 30° to the horizontal and the reading along the inclined limb is 135 mm using the level in the vertical limb as datum. What is the gas pressure?

662 Pa

10.18 For the tank shown in Figure P10.18, determine the pressure in the air space and also the maximum tank pressure (at base).

3.41 kPa; 18.8 kPa

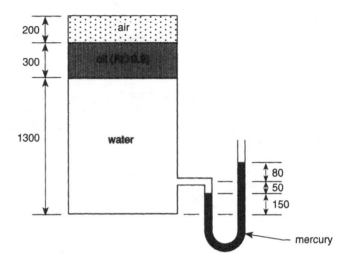

Fig. P10.18

10.19 An inclined piezometer is used to measure water pressure in a horizontal pipe of internal diameter 100 mm. The tube is inclined at 30° to the horizontal and the observed length in the tube is 350 mm (measured from the centre of the pipe). Calculate the water pressure in the pipe. Also calculate the water pressure at the top and bottom of the pipe.
1.72 kPa; 1.23 kPa; 2.21 kPa

10.20 Kerosene flows in a pipe to which is connected a differential manometer, as illustrated in Figure 10.5. Water is used as the indicating fluid and the water level below the centreline of the pipe is 570 mm at ① and 350 mm at ②. Determine the pressure difference between ① and ②. Assume that the relative density of kerosene is 0.78.
475 Pa

10.21 Repeat the calculations for Problem 10.20 if water flows in the pipe and mercury is used as the indicating fluid.
27.2 kPa

10.22 A water tank has a rectangular base of dimensions 3 m × 2 m and vertical sides. When the water level in the tank is 2 m above the base, calculate:
(a) the magnitude and location of the force on the base;
(b) the magnitude and location of the force on the largest side.
    (a) 118 kN; at centre of base   (b) 58.9 kN; 1.33 m below the water surface

10.23 A closed cylindrical storage tank has a circular base section 3.6 m in diameter and a vertical height of 4 m. Calculate the force on the top and bottom of the tank when it is three-quarters full of oil (*RD* 0.9) under a vapour pressure of 10 kPa.
102 kN; 371 kN

10.24 For the tank shown in Figure P10.24 (opposite), determine the tension due to fluid pressure in each of the six bolts holding the cover plate.
505 N

10.25 A sea-water lock is 5 m wide and holds back sea water (*RD* 1.03) to a depth of 3.6 m above the base of the lock. Determine:
(a) the resultant force on the lock;
(b) the turning moment about the base.
    (a) 327 kN   (b) 393 kNm

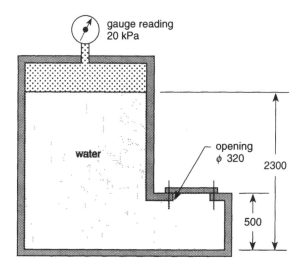

Fig. P10.24

10.26 A vertical freshwater lock is 5 m wide and 3 m high. The water level is 2.7 m above the base of the lock on one side, there being no water on the other side. The mass of the lock is 4 t and the coefficient of friction between the lock and the slides is 0.3. Determine the force necessary to raise the lock.

     92.9 kN

10.27 A horizontal pipe of diameter 600 mm is full of hot water at a pressure of 126 kPa (absolute). The density of the water is 991 kg/m^3 under these conditions. Determine the force on the cover plate that blocks the end of the pipe. Also determine the location of the force relative to the centreline.

     7.81 kN; 7.92 mm below the centreline

10.28 A water trough 2 m wide has one side inclined at 30° to the horizontal. Determine the force on this side when the depth of water in the tank is 1.5 m. Also determine the vertical distance between the force and the free water surface.

     44.1 kN; 1 m down

10.29 A circular vertical opening 400 mm in diameter is cut into the flat side of an oil storage tank and covered by a plate bolted to the side. Determine the force on the plate and its distance below the oil surface when:
(a) the oil surface is just level with the top of the opening;
(b) the oil surface is 1.6 m above the top of the opening.
The relative density of the oil is 0.93.

     (a) 229 N; 250 mm   **(b)** 2.06 kN; 1806 mm

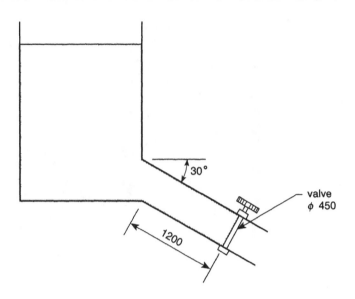

Fig. P10.30

10.30 For the valve shown in Figure P10.30, determine the force on the valve and its location below the centreline of the valve (measured along the valve surface) when the water level is 2 m above the base of the tank.

3.75 kN; 4.56 mm

CHAPTER 11

Fluid flow

Objectives

On completion of this chapter you should be able to:
- describe the meaning of steady flow, streamlines, eddies and average velocity;
- explain the meaning of, and calculate, volume-flow rate and mass-flow rate;
- use the continuity equation in typical fluid-flow applications such as fluid flowing in pipes or ducts with varying cross-sections, with or without branches;
- recognise the Bernoulli equation for flow of an ideal fluid and explain the meaning of 'head' and the various head terms in the equation;
- sketch the variation in the various head terms for typical fluid-flow applications such as fluid flow through tapered and/or inclined pipes;
- explain the causes of head loss that occur with real fluid flow;
- use the Bernoulli equation with or without head losses to determine property changes in typical fluid-flow situations.

Introduction

The previous chapter was concerned with fluid statics, that is, engineering applications where the fluid was at rest or when only the static pressure was considered. This chapter examines the basic principles and equations applicable to fluids in motion.

11.1 BASIC PRINCIPLES

When a fluid flows, the average direction of the fluid can be shown as a series of continuous lines known as **streamlines**. Typical streamlines for flow through a nozzle are shown in Figure 11.1 on page 236. It will be seen that the spacing of the streamlines reflects the velocity and they move closer together when the velocity increases.

The molecules of a fluid are not able to change direction suddenly because this would require an infinite acceleration. Therefore, when there is a sudden change in section as in the case of the sudden contraction shown in Figure 11.1, the streamlines do not follow the sudden change and are rounded. However, both upstream and downstream of the contraction, **eddies** occur where the fluid circulates and actually flows backward in places. In most fluid-flow situations, eddies are undesirable and cause flow losses so it is a general rule that *sudden changes in section or direction should be avoided.*

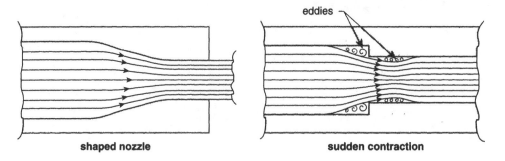

Fig. 11.1   *Streamlines for flow through a shaped nozzle and a sudden contraction*

Steady flow

Steady flow occurs when there is no change to any of the properties of the fluid or to any measured flow variables over a period of time. For example, consider a liquid flowing out of a pipe into a container that has a hole in it as shown in Figure 11.2.

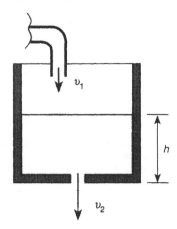

Fig. 11.2   *Steady and unsteady flow of a liquid*

If the liquid flow in the pipe is steady, then the velocity v_1 remains constant. However, if the liquid is flowing into the container at a faster rate than it is flowing out, then both h and v_2 will increase with time, that is, in the container, unsteady-flow conditions exist. However, h and v_2 will not increase indefinitely and after some time, they will not change. From this point onward, there will be steady flow throughout the entire system.

A similar situation occurs in most fluid-flow systems, for example when a pump or fan is first started up. Unsteady flow occurs for a period of time until flow becomes established, after which the system may settle down to a steady-state condition. In this book in all fluid-flow calculations steady flow will be assumed or, at least, any changes to the flow take place gradually so that they do not significantly affect the accuracy of the results obtained by assuming steady flow.

Velocity

All molecules of flowing fluid seldom have the same velocity magnitude or direction and the velocity at any point is known as the local velocity. Even when the fluid flows in a pipe of constant area, the local velocity across a section of the pipe varies and typically is as shown in Figure 11.3.

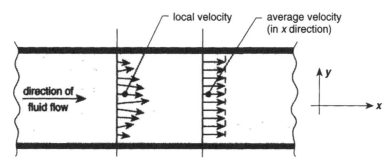

Fig. 11.3 *Local and average velocity*

Because the fluid is flowing from left to right, the majority of the molecules must have a velocity component in the x direction. At any section across the flow path, the average velocity (in the x direction) is *the constant velocity in the x direction all molecules would need to have in order to produce the flow rate resulting from the actual velocity distribution.* This is also depicted in Figure 11.3

In fluid mechanics, 'velocity' (unqualified) always means average velocity. For example if it is stated that the velocity of air in a duct is 10 m/s, then this means the *average velocity* is 10 m/s.

11.2 *VOLUME-FLOW AND MASS-FLOW RATES*

The **volume-flow rate** (or volumetric flow rate) of a fluid is defined as the flow rate expressed as a volume per unit time. This was already treated in Chapter 4 and is given by Equation 4.4:

$$\dot{V} = \frac{V}{t} = \upsilon A \qquad (4.4)$$

The **mass-flow rate** of a fluid is defined as the flow rate expressed as a mass per unit time. This was also treated in Chapter 4 and was given by Equation 4.5:

$$\dot{m} = \frac{m}{t} = \upsilon A \rho \qquad (4.5)$$

Notes

1 In these equations, the dot notation is used to indicate *per unit time*.

2 The following meanings and units apply to the symbols used:

V = volume in m^3 and $\dot{V}$ = volume-flow rate in m^3/s

v = velocity (average) in m/s

t = time in s

A = cross-sectional area (internal) in m^2

m = mass of fluid in kg and $\dot{m}$ = mass-flow rate in kg/s

ρ = density of the fluid in kg/m^3

3 Sometimes volume-flow rate is expressed in L/s or m^3/h. However, base units as quoted above should always be used in Equations 4.4 and 4.5.

4 Chapter 4, Section 4.7 may be referred to for additional information and problems on volume- and mass-flow rates.

Self-test problem 11.1

(a) Water flows through a 10 mm diameter nozzle into a 25 L container, which is filled in 4 minutes 24 seconds. What is the volume-flow rate and the velocity in the nozzle?

(b) Oil of relative density 0.9 flows in a 200 mm inside diameter pipe with velocity 5.6 m/s. What is the volume-flow rate and the mass-flow rate?

11.3 CONTINUITY EQUATION

The continuity equation was given in Chapter 4 and is an application of the conservation of mass principle applied to fluid flow. Equation 4.3 is known as the continuity equation and is:

$$\boxed{\sum \dot{m}_{\text{in}} \ = \ \sum \dot{m}_{\text{out}} \ = \text{constant}}$$

(4.3)

If the density of the fluid does not change appreciably between inlet and outlet, the continuity equation may be written:

$$\boxed{\sum \dot{V}_{\text{in}} \ = \ \sum \dot{V}_{\text{out}} \ = \text{constant}}$$

(4.6)

Notes

1 The assumption of constant density is usually sufficiently accurate with liquids despite changing pressures and temperatures. However, with gases, this assumption is accurate if there are only small changes to the pressure and temperature of the gas.

2 The Σ sign in these equations means 'sum of' and is used because there could be several input flows and several output flows as is the case in Example 11.1.

3 Chapter 4, Section 4.7 may be referred to for additional information and problems on the continuity equation.

Example 11.1

For the pipe shown in Figure 11.4, water enters the pipe at a rate of 15 L/s. Assuming that the velocity in both branches is the same, determine:

(a) the velocity in the pipe;
(b) the velocity in each branch;
(c) the volume-flow rate in each branch.

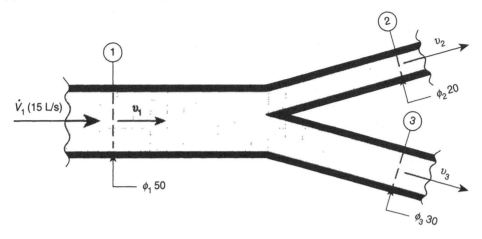

Fig. 11.4

Solution

(a) From Equation 4.6:

$$\dot{V}_1 = v_1 A_1 \text{ and } A_1 = A(0.05) = 1.9635 \times 10^{-3} \text{ m}^3$$
$$\therefore\ 15 \times 10^{-3} = v_1 \times 1.9635 \times 10^{-3}$$
$$\therefore\ v_1 = \textbf{7.64 m/s}$$

(b) Since the fluid is a liquid, the constant-density form of the continuity equation (Equation 4.6) may be used. The equation becomes:

$$\dot{V}_1 = v_2 A_2 + v_3 A_3$$

Since $v_2 = v_3 = v$ (say)

$$\therefore\ v = \frac{\dot{V}_1}{A_2 + A_3}$$
$$A_2 = A(0.02) = 0.3142 \times 10^{-3} \text{ m}^2$$
$$A_3 = A(0.03) = 0.7069 \times 10^{-3} \text{ m}^2$$
$$\therefore\ v = \frac{15 \times 10^{-3}}{0.3142 \times 10^{-3} + 0.7069 \times 10^{-3}}$$
$$= \textbf{14.7 m/s}$$

(c) $\dot{V}_2 = v_2 A_2 = 14.7 \times 0.3142 \times 10^{-3} \text{ m}^3/\text{s} = \textbf{4.62 L/s}$

$\dot{V}_3 = v_3 A_3 = 14.7 \times 0.7069 \times 10^{-3} \text{ m}^3/\text{s} = \textbf{10.38 L/s}$

Check:

$$\dot{V}_1 = \dot{V}_2 + \dot{V}_3$$
$$15 = 4.62 + 10.38$$

which agrees.

Self-test problem 11.2

Steam is produced from a steam generator at a rate of 12.6 t/h. The steam flows along a pipe of diameter 250 mm, which has some leaking joints. At the end of the pipe the steam has specific volume 0.365 m³/kg and velocity 25 m/s. Determine the leakage loss of steam as a percentage of the steam produced.

11.4 *HEAD*

Head is a concept often used in fluid mechanics and is important because head is directly proportional to the energy of a fluid stream.

Head is the height to which a fluid will rise as a result of its energy. There are four types of head, namely pressure head, velocity head, potential head and total head. These are now treated.

Types of heads

Pressure head (h_p)

Pressure head is the height to which a fluid will rise because of its pressure. It is expressed mathematically by changing the subject of Equation 2.8:

$$p = \rho g h$$

∴

$$\boxed{h_p = \frac{p}{\rho g}}$$

pressure head (11.1)

Units: h_p is in m when p is in Pa and ρ is in kg/m³.

Notes

1 Pressure head may be calculated using gauge or absolute pressure. If absolute pressure is used, pressure head must always be positive but if gauge pressures are used, pressure head can be negative.

2 Pressure head of a liquid can be visually seen as height on the static tube of a piezometer as shown in Figure 11.5 (provided it is positive).

Velocity head (h_v)

Velocity head is the height to which a fluid will rise because of its velocity. It is given by:

$$\boxed{h_v = \frac{v^2}{2g}}$$

velocity head (11.2)

Units: h_v is in m when v is in m/s.

Notes

1 Velocity head of a liquid can be visually seen as the difference in height between the dynamic and static tubes of two piezometers placed together as shown in Figure 11.5.

2 This is essentially a Pitot tube as described in Chapter 10, except that the two limbs of a Pitot tube are usually connected to a differential manometer.

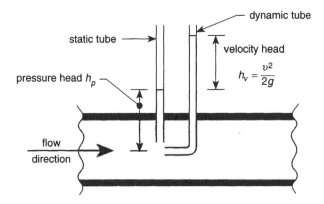

Fig. 11.5 *Pressure and velocity head as shown by piezometers*

Potential head (*h*)

Potential head is the height of the centroid of a fluid stream above a datum. That is, if the fluid is flowing full in pipe, the potential head at any position is the height of the centreline of the pipe at that position above a datum. If the pipe drops below datum, potential head will be negative in this position.

Total head (*H*)

Total head is the sum of the pressure, velocity and potential heads; that is:

$$H = h_p + h_v + h$$ total head (11.3)

Self-test problem 11.3

Oil of relative density 0.9 flows in a pipe with velocity 5 m/s. At a certain point in the pipe the pressure is 30 kPa and the elevation of the pipe is 3.5 m (above datum). Determine the pressure head, velocity head, potential head and total head at this point.

11.5 *BERNOULLI EQUATION*

In the ideal case of a flowing fluid, flow losses and density changes are negligible. If, in addition, there is no fluid machinery (pumps of turbines) between two points in the fluid stream ① and ②, then the total head of the flowing fluid does not change between these points. That is:

$$H = \text{constant or } H_1 = H_2$$

or

$$\frac{p_1}{\rho g} + \frac{v_1^2}{2g} + h_1 = \frac{p_2}{\rho g} + \frac{v_2^2}{2g} + h_2$$ Bernoulli equation (11.4)

Notes

1 This equation is called the Bernoulli equation in honour of Daniel Bernoulli who first formulated it.

2 All the terms in the equation are head terms and all have units m.

3 The continuity equation is often needed in conjunction with the Bernoulli equation and is not implied in it.

4 Pressure may be in Pa gauge or Pa absolute provided consistency is maintained on both sides of the equation.

5 Whereas the Bernoulli equation is a theoretical one since it applies only to an ideal situation, it may be modified to include flow losses as will be done later in this chapter. The equation may also be modified for the case where there is fluid machinery between the points being considered and this modification is treated in the next chapter.

6 In many cases, one or more terms in the equation drop out. Some common cases are listed below:

   - *Horizontal pipe*   There is no change in elevation so the potential head terms drop out on both sides of the equation.
   - *Surface of a liquid at atmospheric pressure*   If gauge pressures are being used in the equation, then pressure head at the liquid surface will be zero.
   - *Fluid discharging through a nozzle to the atmosphere*   If gauge pressures are being used in the equation, then pressure head at the nozzle outlet will be zero.
   - *Fluid flowing through a parallel pipe*   There is no change in velocity so the velocity head terms drop out on both sides of the equation.
   - *Liquid flowing out of, or into a large tank, pond or lake*   At the surface the velocity is negligible and the velocity head at the surface may be taken to be zero.

Pictorial representation of the Bernoulli equation

The various head terms in the Bernoulli equation can be shown diagrammatically. For example, Figure 11.6 (opposite) shows these for the most general case of a fluid flowing through a tapered, inclined tube.

In this case:

potential head:	$h_2 > h_1$	because the tube slopes upward
velocity head:	$h_{v2} > h_{v1}$	because the diameter reduces between ① and ② so $v_2 > v_1$
pressure head:	$h_{p2} < h_{p1}$	because total head is constant and since both potential and velocity heads increase between ① and ②, pressure head reduces

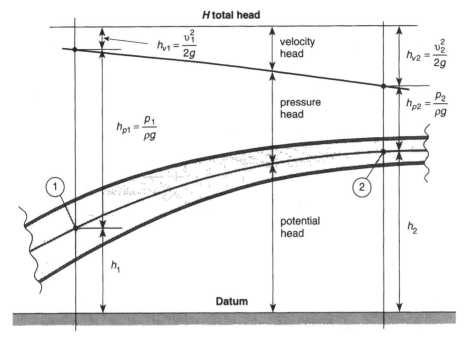

Fig. 11.6

Example 11.2

A horizontal venturi tube inserted in a 75 mm diameter pipe has a throat diameter of 50 mm. If water flows in the pipe such that the upstream pressure and velocity are 45 kPa and 4 m/s respectively, determine the ideal velocity and pressure at the throat.

Solution

$$\phi_1 = 75 \text{ mm}, \ \phi_2 = 50 \text{ mm}$$

$$v_1 = 4 \text{ m/s}, \ v_2 = 4 \times \left(\frac{75}{50}\right)^2 = \textbf{9 m/s}$$

$$h_1 = h_2 \text{ (horizontal pipe)}$$
$$p_1 = 45 \text{ kPa} = 45 \times 10^3 \text{ Pa}$$
$$\rho = 10^3 \text{ kg/m}^3 \text{ (water)}$$

Bernoulli's equation is:

$$\frac{p_1}{\rho g} + \frac{v_1^2}{2g} + h_1 = \frac{p_2}{\rho g} + \frac{v_2^2}{2g} + h_2$$

But $h_1 = h_2$ and multiplying both sides by g:

$$\frac{p_1}{\rho} + \frac{v_1^2}{2} = \frac{p_2}{\rho} + \frac{v_2^2}{2}$$

$$\therefore \ \frac{45 \times 10^3}{10^3} + \frac{4^2}{2} = \frac{p_2}{10^3} + \frac{9^2}{2}$$

$$\therefore \ 45 + 8 = \frac{p_2}{10^3} + 40.5$$

$$\therefore \ p_2 = 12.5 \times 10^3 \text{ Pa} = \textbf{12.5 kPa}$$

Self-test problem 11.4

The pipe shown in Figure 11.7 discharges 40 kg/s of oil (relative density 0.9). The pressure at ② is atmospheric and losses between ① and ② may be neglected.

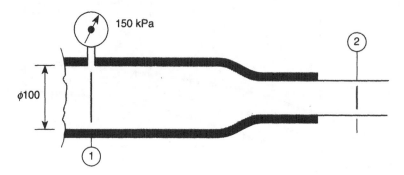

150 kPa

②

φ100

①

Fig. 11.7

Determine:
(a) the velocity at ①;
(b) the velocity head at ①;
(c) the pressure head at ①;
(d) the velocity head at ②;
(e) the velocity at ②;
(f) the diameter of the oil stream at ②.

Example 11.3

A water tank as shown in Figure 11.8 discharges water through an orifice in the side of the tank. Determine the velocity of the water at ② when $h = 2$ m if the air space pressure is **(a)** atmospheric and **(b)** 50 kPa (gauge). Losses may be neglected.

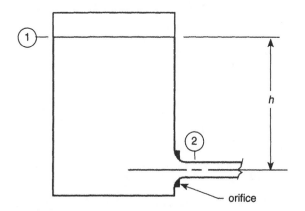

①

②

h

orifice

Fig. 11.8

Solution

(a) Using the centreline of the orifice as datum:

$$p_1 = 0 \text{ (atmospheric)}$$
$$v_1 = 0 \text{ (if the tank is large } v_1 \text{ may be neglected)}$$
$$h_1 = h \text{ (② is datum and } h_2 = 0)$$
$$p_2 = 0 \text{ (atmospheric)}$$

Substituting in Bernoulli's equation:

$$0 + 0 + h = 0 + \frac{v_2^2}{2g} + 0$$

$$\therefore v_2 = \sqrt{2gh}$$

Notes

1 This equation was discovered experimentally by Torricelli (1643) long before the Bernoulli equation was formulated and for this reason is often called the Torricelli equation.

2 A conclusion follows that at first sight may appear illogical, namely that the velocity is independent of the area of the orifice.

3 The equation is the same as for any freely falling body if friction is neglected.

Substituting in the Torricelli equation with $h = 2$ m:

$$v_2 = \sqrt{2 \times 9.81 \times 2}$$
$$= \textbf{6.26 m/s}$$

(b) In this case p_1 is not zero, hence Bernoulli's equation becomes:

$$\frac{p_1}{\rho g} + 0 + h = 0 + \frac{v_2^2}{2g} + 0$$

$$\therefore v_2 = \sqrt{2g\left(h + \frac{p_1}{\rho g}\right)}$$

That is, add the pressure head to the potential head and use this as h in Torricelli's equation.

Since $p_1 = 50$ kPa and $\rho = 10^3$ kg/m³ (water),

$$v_2 = \sqrt{2 \times 9.81\left(2 + \frac{50 \times 10^3}{10^3 \times 9.81}\right)}$$
$$= \textbf{11.8 m/s}$$

Self-test problem 11.5

A siphon as illustrated in Figure 11.9 on page 246 is used to drain water from a tank. The tube diameter is 12 mm and, at point ②, the water stream leaving the tube has a diameter of 10 mm. Elevations (levels) relative to a datum are shown. If losses are neglected, determine:

(a) the velocity at outlet point ②;

(b) the flow rate of water through the siphon in L/s;

(c) the water pressure at the uppermost point ③.

Notes

1 The position of the inlet of the siphon in the tank is irrelevant and the siphon will work provided that the outlet is below the liquid-surface level.

2 To operate the siphon, the liquid must first be made to fill the tube. After this has been done the siphon will work without any external energy being needed.

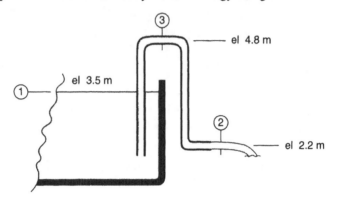

Fig. 11.9

Example 11.4

Water flows in a pipe as shown in Figure 11.10. At ① the diameter is 500 mm, elevation 3 m, pressure 45 kPa and velocity 5.6 m/s. At ② the diameter is 440 mm and the elevation is 5 m. Determine the ideal pressure at ②.

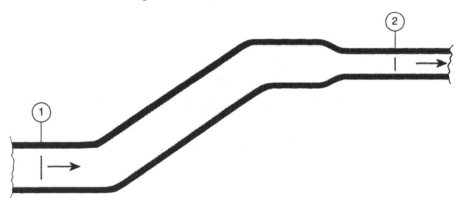

Fig. 11.10

Solution

Data:

$$\phi_1 = 500 \text{ mm} \qquad \phi_2 = 440 \text{ mm}$$
$$h_1 = 3 \text{ m} \qquad h_2 = 5 \text{ m}$$
$$p_1 = 45 \text{ kPa} \qquad \rho = 10^3 \text{ kg/m}^3 \text{ (water)}$$
$$v_1 = 5.6 \text{ m/s}$$

From continuity,

$$v_2 = v_1 \left(\frac{\phi_1}{\phi_2}\right)^2 = 5.6 \times \left(\frac{500}{440}\right)^2 = 7.23 \text{ m/s}$$

Bernoulli's equation is:

$$\frac{p_1}{\rho g} + \frac{v_1^2}{2g} + h_1 = \frac{p_2}{\rho g} + \frac{v_2^2}{2g} + h_2$$

Substituting:

$$\frac{45 \times 10^3}{10^3 \times 9.81} + \frac{5.6^2}{2 \times 9.81} + 3 = \frac{p_2}{10^3 \times 9.81} + \frac{7.23^2}{2 \times 9.81} + 5$$

$$\therefore 4.587 + 1.598 + 3 = \frac{p_2}{10^3 \times 9.81} + 2.665 + 5$$

$$\therefore \frac{p_2}{10^3 \times 9.81} = 1.52$$

$$\therefore p_2 = \textbf{14.9 kPa}$$

Self-test problem 11.6

Liquid with relative density 1.15 flows at a rate of 9.5 kg/s through the pipe system shown in Figure 11.11. The pressure at point ② is 30 kPa (gauge).

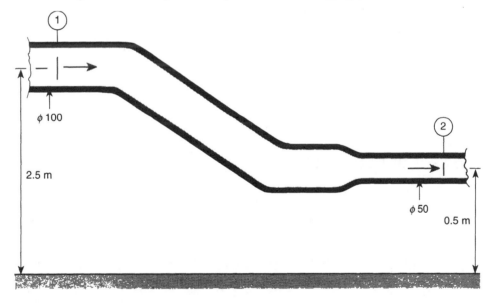

Fig. 11.11

Neglecting losses, for points ① and ② determine://
(a) the potential head;
(b) the velocity head;
(c) the pressure head.
(d) Also draw a neat sketch showing how these heads vary between ① and ② and mark values at these points on the sketch.

11.6 *HEAD LOSS*

So far, the Bernoulli equation has been treated for ideal flow with no fluid viscosity and no friction loss. As was seen in Chapter 8, every fluid (whether a liquid or a gas) has viscosity and therefore a flow loss due to fluid friction will occur. As well as this, non-streamline-flow conditions (such as sharp corners) will cause eddies and turbulence, which will increase the flow loss. The flow loss causes a **head loss**, which means that the total head of the fluid does not remain constant but reduces as shown in Figure 11.12 (which is Figure 11.6 redrawn to include head loss).

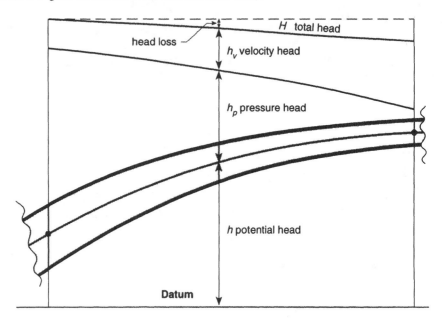

Fig. 11.12   *Flow loss reduces the total head*

Effect of head loss

What effect occurs as a result of head loss? To answer this question, refer to Figure 11.12 and assume that the initial conditions (point ①) are the same for both the ideal and the real fluid. What changes to the real fluid compared with the ideal will occur downstream (point ②)?

- *Total head*   The total head at ② will be less due to the head loss. That is:

$$H_2 = H_1 - H_L \quad \text{or} \quad H_1 = H_2 + H_L$$

- *Potential head*   The potential head at ② will not change because this depends on the elevation of the fluid stream, which is not affected by head loss.
- *Velocity head*   The velocity head at ② will not change if the fluid is fully constrained (as is the case if the fluid is running full in a pipe or duct) because the conservation of mass law (equation of continuity) cannot be violated and is true for all types of fluids whether real or ideal. Therefore, for given upstream conditions (point ①), the flow rate, velocity and velocity head downstream at point ② will not change, regardless as to the head loss.

However, if the fluid is not constrained (as is the case with a free jet of fluid), the velocity will reduce because of friction. When this occurs, the jet spreads out because as the velocity reduces, the cross-sectional area must increase so as not to violate the equation of continuity. In this case the pressure head does not change because the fluid is at constant pressure (atmospheric).

- *Pressure head* If both the potential head and the velocity head do not change, the pressure head at ② must reduce. This is shown in Figure 11.12. Therefore, when a fluid flows full in a duct or pipe, the head loss manifests itself as a *reduction in pressure head and therefore in the pressure of the fluid*. This reduction in pressure caused by a head loss may readily be calculated by modifying Equation 2.8:

$$p = \rho g h$$

Therefore

$$\boxed{\Delta p = \rho g H_L}$$ pressure drop due to head loss (11.5)

Where Δp is the pressure drop ($p_1 - p_2$) in Pa and H_L is the head loss in m.

Modification of the Bernoulli equation to include head loss

The Bernoulli equation is readily modified to include head loss so the equation is valid for real fluid flow and not just ideal fluid flow. For real fluids:

$H_1 = H_2 + H_L$. Therefore:

$$\boxed{\frac{p_1}{\rho g} + \frac{v_1^2}{2g} + h_1 = \frac{p_2}{\rho g} + \frac{v_2^2}{2g} + h_2 + H_L}$$ Bernoulli equation with head loss (11.6)

Example 11.5

Liquid, *RD* 0.8 flows with velocity 4 m/s in a pipe that has a downward slope of 1:50 (sine). At a certain point in the pipe, a pressure gauge shows a pressure of 80 kPa. Determine the pressure at a point 200 m downstream of the gauge if:
(a) flow losses are ignored (ideal case);
(b) there is a flow loss equal to 10% of the total initial head.

Solution

(a) Using the downstream point as datum, the potential head $h_2 = 0$ and because the pipe is a parallel one, the velocity head does not change. Therefore the Bernoulli equation for ideal flow becomes:

$$\frac{p_1}{\rho g} + \frac{v_1^2}{2g} + h_1 = \frac{p_2}{\rho g} + \frac{v_2^2}{2g} + h_2$$

$$\therefore \frac{80 \times 10^3}{0.8 \times 10^3 \times 9.81} + \frac{200}{50} = \frac{p_2}{\rho g}$$

$$\therefore \frac{p_2}{\rho g} = 14.19$$

$$\therefore p_2 = 14.19 \times 0.8 \times 10^3 \times 9.81 \text{ Pa} = \textbf{111.4 kPa}$$

(b) The velocity head at ① is:

$$h_{v1} = \frac{v_1^2}{2g} = \frac{4^2}{19.62} = 0.816 \text{ m}$$

∴ Total head at ① $= 14.19 + 0.816 = 15$ m

∴ $H_L = 10\% \times 15 = 1.5$ m

Using equation 11.5: $\Delta p = \rho g H_L$

∴ $\Delta p = 0.8 \times 10^3 \times 9.81 \times 1.5$ Pa $= 11.8$ kPa

∴ $p_2 = 111.4 - 11.8 =$ **99.6 kPa**

Self-test problem 11.7

A water blaster used for cleaning purposes has a pump with the outlet pipe 12 mm in diameter. The pipe is connected to a flexible hose at the end of which is a nozzle that discharges a water jet of diameter 5 mm with velocity 50 m/s. Determine the pressure developed at the outlet pipe when the nozzle is 1 m above it if:
(a) losses are neglected;
(b) there is a head loss of 10 m between the outlet of the pump and the water jet.

Summary

A flowing fluid is unable to change direction suddenly and, if there are abrupt changes in section in the flow path such as sharp corners, the average direction of the fluid may be represented by a series of continuous lines known as streamlines with eddies occurring in the region of the voids. If flow conditions do not change appreciably with time, the flow is steady and, under steady-state conditions, the mass of fluid entering and leaving the system must be the same. The mathematical expression of this is known as the equation of continuity. A special case of this equation occurs when the density of the fluid also does not change appreciably, in which case the volume-flow rate into and out of the system must be the same.

Head is a very important concept in fluid mechanics because head is directly proportional to the energy of the fluid. A flowing fluid has velocity head, if under pressure it also has pressure head and if elevated above a datum it also has potential head. The sum of these three heads is the total head. If there are no flow losses, fluid energy and therefore head must be conserved. The mathematical statement of this is known as Bernoulli's equation.

When real fluids flow, a loss of useful energy occurs as a result of friction and eddies. This causes a flow loss so that the total head downstream is less than the total head upstream. That is:

$$H_2 = H_1 - H_L$$

The Bernoulli equation may be modified for real-flow conditions by including a head-loss term in the equation, that is:

$$H_1 = H_2 + H_L$$

The magnitude of the head loss may be calculated but this is beyond the scope of this book.

Problems

In all cases assume dimensions are internal ones and that flow is steady.

11.1 Water flows with velocity 3 m/s in a pipe of diameter 100 mm into which a nozzle of diameter 50 mm is fitted. Determine the velocity of water leaving the nozzle.

    12 m/s

11.2 A water pipe connects to a tank of capacity 50 m³. If the tank is to be filled in one hour, determine the pipe diameter necessary if the velocity in the pipe is not to exceed 5 m/s.

    59.5 mm

11.3 Water flows with velocity 5 m/s through a pipe of diameter 45 mm, which reduces downstream to 30 mm diameter. Determine the velocity in the pipe downstream of the reducer.

    11.25 m/s

11.4 Determine the discharge in kg/s and L/s when oil of relative density 0.92 flows in a pipe of diameter 50 mm with velocity 6 m/s.

    10.8 kg/s; 11.8 L/s

11.5 A sea-water-pumping station is to pump 15.6 t/h of sea water (*RD* 1.03). If it is desired that the velocity in the inlet pipe to the pump is not to exceed 3 m/s and the velocity in the outlet pipe is not to exceed 5 m/s determine suitable inlet and outlet pipe diameters.

    42.3 mm; 32.7 mm

11.6 A sprinkler head consists of a number of holes of diameter 1 mm. If the sprinkler is to discharge 1.5 L/s of water with velocity 30 m/s through the holes, determine the number of holes required.

    64

11.7 A pipe discharges 8 t/h of oil (*RD* 0.9) through a 50 mm diameter pipe into a circular storage tank of diameter 2.3 m. Determine:
(a) the velocity in the 50 mm pipe;
(b) the rate per hour at which the level is rising in the tank;
(c) the time taken to fill the tank to a depth of 5 m.

    (a) 1.26 m/s   (b) 2.14 m/h   (c) 2.34 h

11.8 A pipe of diameter 80 mm is fitted with a branch section into which fit two pipes of equal diameter. If it is desired that the velocity remain constant throughout, determine the required diameter of each of the branch pipes.

    56.6 mm

11.9 A heat exchanger has 30 copper tubes of diameter 25 mm connected in parallel to an inlet and outlet pipe. Oil of relative density 0.85 flows through the exchanger with velocity 3 m/s in both pipe and tubes. Determine:
(a) the volume-flow rate;
(b) the mass-flow rate;
(c) the diameter of the inlet and outlet pipes.

    (a) 44.2 L/s   (b) 37.6 kg/s   (c) 137 mm

11.10 Air (specific volume 0.865 m³/kg) flows at 8 m/s through a circular duct of diameter 300 mm. A branch section of 150 mm diameter is fitted to the duct and the velocity of the air downstream of the branch is 6 m/s (in the 300 mm duct). Determine:

(a) the volume-flow rate;
(b) the mass-flow rate;
(c) the velocity in the 150 mm branch section.
The density of the air may be assumed to be constant.
  (a) 0.566 m^3/s (b) 0.654 kg/s (c) 8 m/s

11.11 Water flows with velocity 5 m/s in a pipe of diameter 300 mm. Further downstream a pipe of diameter 100 mm connects into the 300 mm pipe. The water leaves the 100 mm pipe with velocity 8 m/s. Calculate:
(a) the volume-flow rate;
(b) the mass-flow rate;
(c) the velocity in the 300 mm pipe downstream of the junction.
  (a) 0.353 m^3/s (b) 353 kg/s (c) 4.11 m/s

11.12 A pipe of diameter 200 mm carrying water with velocity 4.43 m/s has an upward slope of 1 : 50 (sine). At a certain point, the elevation of the pipe is 3 m (above a datum) and the pressure is 50 kPa (gauge).
(a) Determine the pressure head, velocity head, potential head and total head at this position.
(b) Recalculate these heads for a position 100 m further up the pipe if losses are neglected.
(c) Recalculate the heads for part (b) if a flow loss of 1.5 m occurs.
  (a) 5.1 m; 1.0 m; 3.0 m; 9.1 m   (b) 3.1 m; 1.0 m; 5.0 m; 9.1 m
  (c) 1.6 m; 1.0 m; 5.0 m; 7.6 m

11.13 A horizontal pipe of diameter 150 mm is fitted with a nozzle of smaller diameter at the end of it. The nozzle discharges 60 kg/s of oil (*RD* 0.89) at atmospheric pressure to the space above an oil tank. At a position in the pipe upstream of the nozzle, the pressure is 120 kPa (gauge).
(a) Determine the pressure head and velocity head at this position.
(b) If flow losses are ignored, determine the velocity and diameter of the oil stream leaving the nozzle.
(c) If a head loss equal to 10% of the total head at (a) occurs, determine what the velocity and diameter of the oil stream leaving the nozzle will now be.
  (a) 13.7 m; 0.742 m   (b) 16.9 m/s; 71.4 mm   (c) 16 m/s; 73.3 mm

11.14 A horizontal pipe 50 mm in diameter is fitted with a venturi tube, which has a throat diameter of 40 mm. The pipe carries water at a pressure of 30 kPa gauge and velocity 3 m/s upstream of the throat. Determine:
(a) the mass-flow rate;
(b) the velocity at the throat;
(c) the pressure at the throat ignoring head losses;
(d) the pressure at the throat if a head loss equal to 5% of the total head occurs as a result of the venturi.
  (a) 5.89 kg/s (b) 4.69 m/s (c) 23.5 kPa (d) 21.8 kPa

11.15 Water discharges from a tank in which the surface level is 3.5 m above the centre of an orifice located in the side of the tank. Determine the ideal velocity of the water jet when:
(a) air-space pressure is atmospheric;
(b) air-space pressure is 20 kPa (gauge);
(c) air-space pressure is –20 kPa (vacuum).
  (a) 8.29 m/s (b) 10.4 m/s (c) 5.35 m/s

11.16 Repeat the calculations in Problem 11.15 for kerosene (*RD* 0.78) instead of water.
(a) 8.29 m/s (b) 11 m/s (c) 4.17 m/s

11.17 Repeat the calculations in Problem 11.15 if a flow loss equal to 6% of the total head occurs between the surface and the outlet of the nozzle.
(a) 8.03 m/s (b) 10.1 m/s (c) 5.19 m/s

11.18 Salt water (*RD* 1.04) is pumped at a rate of 27.5 L/s through a 100 mm diameter pipe from an open pond. The inlet of the pump is 2.6 m above the surface of the pond and the head loss in the suction line is 1.8 m. Determine the gauge pressure at the inlet to the pump.
−51.3 kPa

11.19 Liquid (*RD* 0.9) is pumped through a pipe fitted with a reducing bend as shown in Figure P11.19. At position ① the pressure is 40 kPa (gauge) and the velocity is 5 m/s.

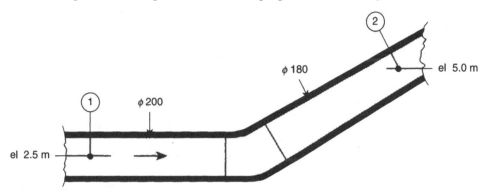

Fig. P11.19

If losses are neglected:
(a) determine the velocity and pressure at ②;
(b) draw a diagram to scale showing the variation in potential head, pressure head, velocity head and total head between ① and ②. Clearly mark the values at ① and ② on the diagram.
(c) If a head loss of 2 m occurs between ① and ②, redraw the diagram for part (b).
(a) 6.17 m/s; 12 kPa (b) ① 2.5 m; 4.53 m; 1.27 m; 8.3 m; ② 5.0 m; 1.36 m; 1.94 m; 8.3 m (c) ① same as (b); ② 5.0 m; −0.64 m; 1.94 m; 6.3 m

11.20 Air flows through a horizontal duct of dimensions 300 × 300 mm with velocity 15 m/s. At position ① in this duct a water gauge (water manometer) registers a height of 215 mm. The duct bends downwards and reduces in size to 240 × 240 mm, dropping a distance of 12 m to position ②. The specific volume of the air is 0.85 m³/kg and this may be taken to be constant. Determine:
(a) the volume-flow rate and mass-flow rate;
(b) the pressure at ① in kPa (gauge);
(c) the velocity at ②;
(d) the pressure at ② in kPa (gauge) and mm water gauge if losses are neglected;
(e) the pressure at ② in kPa (gauge) and mm water gauge if losses are 10% of the total head at ① and elevation ② is used as datum.
(a) 1.35 m³/s; 1.59 kg/s (b) 2.11 kPa (c) 23.4 m/s (d) 2.06 kPa; 210 mm
(e) 1.82 kPa; 185 mm

CHAPTER 12

Fluid power

≡

Objectives

On completion of this chapter you should be able to:

- write the equations for fluid power in terms of mass-flow rate and head and also volume-flow rate and pressure and use these equations in typical engineering applications;

- calculate the useful fluid power lost as a result of a head loss;

- write equations for the efficiency of fluid machinery (pumps or turbines) and use these equations in calculations involving fluid machinery;

- modify Bernoulli's equation to include fluid machinery in the flow path as well as a head-loss term and use the equation in typical engineering applications.

Introduction

Energy and power are associated with the flow of a fluid. The process of conversion from fluid power to mechanical power or vice versa is widely used in engineering. In turbines and motors, fluid power is converted into mechanical power whereas in pumps, fans and compressors, mechanical power is converted into fluid power. This chapter is devoted to this important application of fluid mechanics.

12.1 *FLUID POWER AND HEAD*

It was seen in Chapter 11 that the head of a flowing fluid is directly related to the energy or power of the fluid. A simple important equation expresses this relationship mathematically:

$$\boxed{P_f = \dot{m}\,gH}$$

fluid power (12.1)

Where: P_f = fluid power in W
$\dot{m}$ = mass-flow rate of the fluid in kg/s
g = gravitational constant (9.81 N/kg or m/s²)
H = head of the fluid in m

Notes

1 In this equation, H is the head of the fluid and in the most general application will be the total head. However, in some cases, only one head changes significantly and then the particular head of significance is used. H may also be a head loss, H_L, in which case P_f will be the power associated with this head loss.

2 Fluid power P_f is identical with indicated power IP used with heat engines as defined in Chapter 7. The term 'indicated power' is used with heat engines rather than 'fluid power' because of established usage as, historically, indicated power was measured with an engine indicator.

3 Fluid power may increase or decrease the energy of a fluid stream. For example, pumps increase the fluid power by adding energy to the fluid whereas turbines decrease the fluid power by taking energy out of the fluid stream. Head losses decrease the fluid power by taking energy out of the fluid stream.*

Example 12.1

A turbine used for hydroelectric-power generation has a total head of 86 m at the turbine inlet and 4.6 m at the turbine outlet. If 1.7 m³/s of water flow through the turbine, calculate the change in fluid power.

Solution

In this case $\dot{m} = 1.7 \times 10^3$ kg/s and the head change $H = 86 - 4.6 = 81.4$ m

Using equation 12.1,

$$P_f = \dot{m} gH$$
$$\therefore P_f = 1.7 \times 10^3 \times 9.81 \times 81.4 \quad \text{W}$$
$$= \textbf{1.36 MW}$$

 ## Self-test problem 12.1

A pump is used to pump liquid of relative density 1.1 through a pipe of diameter 65 mm with velocity 2.2 m/s. The head at the inlet to the pump is 3.6 m and at the outlet to the pump is 26.3 m. Determine the change in fluid power due to the pump.

12.2 *FLUID POWER AND PRESSURE HEAD*

In many applications (for example in hydraulics and pneumatics), the fluid power is primarily due to a change in pressure (pressure head). In such cases, Equation 12.1 may be used, with H being the pressure head, h_p. This equation may be modified in terms of pressure rather than pressure head as shown below:

If the head is only pressure head, then $H = \dfrac{p}{\rho g}$

And $P_f = \dot{m}gH = \dfrac{\dot{m}gp}{\rho g} = \dfrac{\dot{m}p}{\rho}$

But $\dot{m} = vA\rho$

---

* Strictly speaking, head losses do not decrease the *total* energy of the fluid, but the *useful* energy. Head losses cause the fluid to heat up slightly and thereby increase its temperature and internal energy but the increase is usually very small and is quickly dissipated to the surroundings.

$$\therefore P = \frac{vA\rho p}{\rho} = vAp$$

But $vA = \dot{V}$ (volume-flow rate)

Therefore

$$\boxed{P_f = p\dot{V}}$$ fluid power due to a pressure change (12.2)

Where P_f = fluid power in W

p = pressure change in Pa

$\dot{V}$ = volume-flow rate in m³/s

Example 12.2

In a hydraulic application, hydraulic oil of relative density 0.82 is pumped through an 8 mm diameter pipe with velocity 2.5 m/s. The pressure at the inlet to the pump is atmospheric and at the outlet to the pump is 6.5 MPa (abs). Determine the fluid-power change due to the pump using both Equations 12.1 and 12.2 if changes to potential and velocity head are negligible.

Solution

$$\dot{V} = vA = 2.5 \times A(0.008) = 0.1257 \times 10^{-3} \text{ m}^3/\text{s}$$

$$\dot{m} = vA\rho = \dot{V}\rho = 0.1257 \times 10^{-3} \times 0.82 \times 10^3 = 0.103 \text{ kg/s}$$

The change in pressure caused by the pump, $p = 6500 - 101.3 = 6398.7$ kPa

$$H = h_p = \frac{p}{\rho g} = \frac{6398.7 \times 10^3}{820 \times 9.81} = 795.4 \text{ m}$$

Using Equation 12.1: $P_f = \dot{m}gH = 0.103 \times 9.81 \times 795.4 = \textbf{804 W}$

Using Equation 12.2: $P_f = p\dot{V} = 6398.7 \times 10^3 \times 0.1257 \times 10^{-3} = \textbf{804 W}$

 ## Self-test problem 12.2

Air with velocity 8.3 m/s flows through a fan in a circular duct of diameter 500 mm. The air pressure at the inlet to the fan is 20 mm water gauge and at the outlet to the fan is 126 mm water gauge. The specific volume of the air is 0.83 m³/kg and may be taken to be constant. Determine:

(a) the volume-flow rate;
(b) the mass-flow rate;
(c) the change in pressure due to the fan in Pa;
(d) the change in head due to the fan in m;
(e) the change in fluid power using Equation 12.1;
(f) the change in fluid power using Equation 12.2.

12.3 *FLUID POWER AND HEAD LOSS*

The loss of useful power resulting from a head loss may be calculated using Equation 12.1 with $H = H_L$. Because head loss is all in pressure head, Equation 12.2 may also be used with p = pressure drop due to the head loss.

Self-test problem 12.3

A salt solution with relative density 1.05 is pumped through a horizontal pipe of diameter 250 mm with velocity 2.8 m/s. A mercury manometer is connected to the upstream end and reads 284 mm. When the manometer is connected to the downstream end, so that the interface level is the same distance below the pipe, the reading is 275 mm. The relative density of mercury is 13.6.

Determine:

(a) the pressure drop in kPa;

(b) the head loss in m;

(c) the fluid power needed to overcome frictional resistance in this length of pipe using both Equations 12.1 and 12.2.

12.4 *EFFICIENCY OF FLUID MACHINERY*

As was discussed in Chapter 9, fluid machines convert mechanical power into fluid power (pumps or compressors) or fluid power into mechanical power (turbines and motors). The ratio between the mechanical and fluid powers is the efficiency and is therefore given by:

$$\eta = \frac{P_f}{P}$$ efficiency of a pump or compressor (12.3)

$$\eta = \frac{P}{P_f}$$ efficiency of a turbine or motor (12.4)

Notes

1 In these Equations, P is the mechanical power. For linear movement, mechanical power is given by Equation 3.8: $P = Fv$ and for rotational movement is given by Equation 3.9: $P = T\omega$.

2 Efficiency is usually expressed as a percentage and is always less than 100% due to losses such as friction, eddies, backflow and leakage.

3 Equation 12.4 is identical with Equation 7.5 for a heat engine remembering that $P_f = IP$ for a heat engine.

Example 12.3

A water turbine rotating 600 rpm produces 450 kW of output power when the flow rate of water is 1.95 m^3/s and the change in fluid head across the turbine is 25 m. Determine:

(a) the efficiency of the turbine;

(b) the output torque.

Solution

(a) $P_f = \dot{m}gH = 1950 \times 9.81 \times 25$ W $= 478$ kW

$$\eta = \frac{P}{P_f} = \frac{450}{478} = 0.941 \text{ or } \mathbf{94.1\%}$$

(b) $$P = T\omega = T \times \frac{\pi N}{30}$$

$$\therefore 450 \times 10^3 = T \times \frac{\pi \times 600}{30}$$

$$\therefore T = \mathbf{7.16 \text{ kNm}}$$

 ## Self-test problem 12.4

The inlet and outlet flanges of a water pump are 100 mm in diameter and are at the same horizontal level. The velocity of water entering (and leaving) the pump is 3.2 m/s and the pressure at the inlet flange is –22 kPa (gauge). The input power is 15 kW and the efficiency is 65%. Determine the pressure at the outlet flange of the pump.

12.5 *BERNOULLI EQUATION WITH FLUID MACHINERY*

In Chapter 11, Bernoulli's equation was treated for fluid-flow situations where there was no fluid machinery between the two positions ① and ② in the flow path. Consider now the case where fluid machinery is installed in the flow path between the two positions as shown in Figure 12.1.

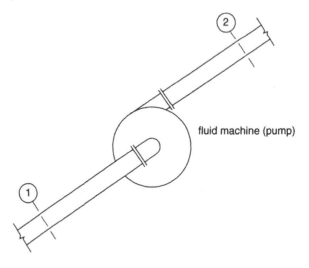

fluid machine (pump)

Fig. 12.1   *Fluid machinery in the flow path*

If the head change caused by the fluid machine is H and there is a head loss H_L between points ① and ②, then:

$$H_1 \pm H = H_2 + H_L$$

The plus or minus symbol (±) is necessary because the fluid head may be increased or decreased depending on the type of fluid machine. For example, if the machine is a pump, fluid head will increase (+) but for a turbine or motor, fluid head will decrease (–).

In the same way, the full Bernoulli equation (Equation 11.6) becomes:

$$\frac{p_1}{\rho g} + \frac{v_1^2}{2g} + h_1 \pm H = \frac{p_2}{\rho g} + \frac{v_2^2}{2g} + h_2 + H_L$$

Bernoulli equation with fluid machinery (12.5)

Example 12.4

A pumping system as shown in Figure 12.1 is used to pump water at a rate of 30 L/s. The following information is given:

	①	②
pressure (kPa)	25	80
pipe diameter (mm)	100	75
elevation (m)	3	9

Determine the power input to the pump if its efficiency is 60%,
(a) neglecting losses between ① and ②;
(b) if the head loss between ① and ② is 6 m.

Solution

$$v_1 = \frac{\dot{V}}{A_1} = \frac{30 \times 10^{-3}}{A(0.1)} = 3.82 \text{ m/s}$$

$$v_2 = \frac{\dot{V}}{A_2} = \frac{30 \times 10^{-3}}{A(0.075)} = 6.79 \text{ m/s}$$

(a) If losses are neglected, $H_L = 0$, also H is positive (pump). Equation 12.5 becomes:

$$\frac{p_1}{\rho g} + \frac{v_1^2}{2g} + h_1 + H = \frac{p_2}{\rho g} + \frac{v_2^2}{2g} + h_2$$

Substituting:

$$\frac{25 \times 10^3}{10^3 \times 9.81} + \frac{3.82^2}{2 \times 9.81} + 3 + H = \frac{80 \times 10^3}{10^3 \times 9.81} + \frac{6.79^2}{2 \times 9.81} + 9$$

$$2.55 + 0.744 + 3 + H = 8.15 + 2.35 + 9$$
$$\therefore H = 13.2 \text{ m}$$

Now $P_f = \dot{m}gH$
$$\therefore P_f = 30 \times 9.81 \times 13.2 \quad \text{W}$$
$$= 3.89 \text{ kW}$$

which is the fluid power of the pump.

Since for a pump $\eta = \dfrac{\text{fluid power}}{\text{input power}} = \dfrac{P_f}{P}$

input power $P = \dfrac{3.89}{0.6} = \textbf{6.48 kW}$

(b) If losses are included the head-loss term must be included in the right-hand side of Equation 12.5.

$$\therefore H = 13.2 + 6 = 19.2 \text{ m}$$
$$\therefore P = 30 \times 9.81 \times 19.2 = 5.65 \text{ kW}$$

$$\therefore \text{Input power} = \frac{5.65}{0.6} = \textbf{9.42 kW}$$

 ## Self-test problem 12.5

A water turbine as shown in Figure 12.2 is 92% efficient. The head loss in the piping system is 0.6 m. Determine the power produced by the turbine per m³ of water flowing through it per second.

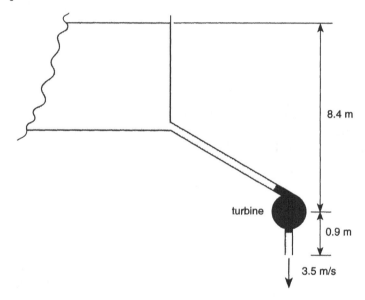

8.4 m

turbine

0.9 m

3.5 m/s

Fig. 12.2

 # Summary

The relationship between fluid power and head may be expressed by a single equation that holds in all cases, namely: $P_f = \dot{m}gH$. This equation may be used for fluid machinery and also to calculate the power associated with a head loss by using H_L in place of H. In many engineering applications, only the pressure head is of significance, in which case the equation may be written in terms of pressure as: $P_f = p\dot{V}$.

Fluid machines in the form of turbines or motors convert fluid power to mechanical power by extracting energy from the fluid stream. Their efficiency is given by $\eta = \dfrac{P}{P_f}$.

Fluid machines in the form of pumps, fans and compressors convert mechanical power to fluid power by adding energy to the fluid stream. Their efficiency is given by:

$$\eta = \frac{P_f}{P}.$$

Bernoulli's equation may be modified to include the case where a fluid machine is located in the fluid-flow path between the two positions being considered. In this case:

$$H_1 \pm H = H_2 + H_L$$

The full equation is:

$$\frac{p_1}{\rho g} + \frac{v_1^2}{2g} + h_1 \pm H = \frac{p_2}{\rho g} + \frac{v_2^2}{2g} + h_2 + H_L$$

If energy is added to the fluid stream (pump or compressor) H is positive, whereas when energy is extracted from the fluid stream (fluid motor or turbine), H is negative.

Problems

12.1 The fluid power transferred by a pump to water flowing in a horizontal pipe at rate of 40 L/s is 8 kW. Determine:
(a) the increase in fluid head and state which fluid head is involved;
(b) the pressure at the pump outlet if the pressure at the pump inlet is –10 kPa;
(c) the efficiency of the pump if the input power is 12.6 kW.
  (a) 20.4 m (pressure head)   (b) 190 kPa   **(c)** 63.5%

12.2 Brine (*RD* 1.08) is pumped at a rate of 20 L/s in a horizontal pipe. The pressure increase across the pump is 250 kPa. Determine:
(a) the change in head across the pump;
(b) the increase in fluid power;
(c) the input shaft power if the efficiency of the pump is 80%.
  (a) 23.6 m   (b) 5 kW   (c) 6.25 kW

12.3 Determine the fluid power necessary to produce a free supersonic water jet of diameter 100 mm and velocity 800 m/s (about Mach 2.4) from a reservoir of water at atmospheric pressure. Comment on the answer.
  2011 MW. This is an enormous amount of power and is equivalent to the combined power output of two large coal fired-power stations!

12.4 A pump is used to transfer 20 L/s of water from a reservoir at elevation 3 m to another at elevation 11 m. Determine:
(a) the fluid power required if there are no losses;
(b) the fluid power required if the actual head loss is 1.3 m;
(c) the actual power required to drive the pump if the efficiency of the pump is 62%.
  (a) 1.57 kW   (b) 1.82 kW   (c) 2.94 kW

12.5 A wind-driven propeller of diameter 10 m is coupled to an electric generator and has an overall efficiency of 45%. The average specific volume of air is 0.85 m³/kg and the wind speed upstream of the propeller is 21 km/h, downstream of the propeller is 7 km/h and at the propeller itself is the mean of these values, namely 14 km/h. Determine:
(a) the mass-flow rate of air through the propeller;
(b) the change in head of the air passing through the propeller (note that this is all a change in velocity head);
(c) the change in fluid power caused by the propeller;

(d) the electrical power output.

   (a) 359 kg/s   (b) 1.54 m   (c) 5.43 kW   (d) 2.45 kW

12.6 A water turbine (Pelton wheel) has a series of buckets on a pitch diameter of 1.2 m. The wheel turns at 240 rpm and the average force on the buckets is 70 N. The turbine is driven by a water jet discharging 2 kg/s of water at atmospheric pressure into the buckets with velocity 36 m/s. Calculate:
   (a) the fluid power;
   (b) the output power;
   (c) the turbine efficiency.

   (a) 1.296 kW   (b) 1.056 kW   (c) 81.5%

12.7 The centrifugal pump illustrated in Figure P12.7 delivers 25 L/s of water with a total head increase of 15 m. Determine:
   (a) the potential head change;
   (b) the velocity head change;
   (c) the pressure head change;
   (d) the fluid power;
   (e) the input mechanical power if the efficiency is 70%;
   (f) the cost of operating the pump per hour if it is driven by an electric motor that has an efficiency of 85% and electrical energy costs 12 cents per kWh.

   (a) 0.23 m   (b) 0.744 m   **(c)** 14.03 m   (d) 3.68 kW   (e) 5.26 kW   (f) 74.2 cents

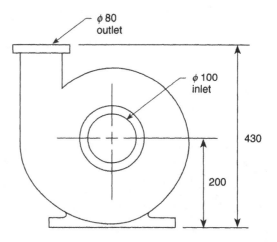

Fig. P12.7

12.8 A hydraulic motor operates on an inlet-oil pressure of 5.5 MPa and an outlet-oil pressure of 500 kPa. The motor has an efficiency of 70% and the output power is 5 kW at a speed of 500 rpm. The relative density of the oil is 0.9 and changes in potential and kinetic energy are negligible. Determine:
   (a) the flow rate of oil in L/s;
   (b) the minimum diameter of the inlet and outlet ports so that the oil velocity does not exceed 4 m/s;
   (c) the output torque.

   (a) 1.43 L/s   (b) 21.3 mm   (c) 95.5 Nm

12.9 A pump transfers 8 kg/s of liquid with relative density 0.75 from an open tank into a closed tank. The liquid elevation in the open tank is 2.8 m above datum and in the closed tank 5.3 m. Determine:

(a) the theoretical (loss-free) power required if the closed tank is vented to atmosphere;

(b) the theoretical (loss-free) power required if the closed tank is under a vapour pressure of 120 kPa (gauge);

(c) the actual power required if the closed tank is under a vapour pressure of 120 kPa (gauge) and the head loss is 9 m.

    (a) 196 W   (b) 1.48 kW   (c) 2.18 kW

12.10 Oil (*RD* 0.9) is pumped at a rate of 25 L/s from a tank under a vapour pressure of 30 kPa (gauge). The discharge of the pump is connected to a pipe of diameter 100 mm and the pump imparts 5 kW of fluid power to the oil. Determine the theoretical (loss-free) pressure that would be shown by a gauge connected to the discharge pipeline at an elevation of 8 m above the oil level in the tank.

    Also determine the actual pressure shown by the gauge if the head loss between the tank and the gauge is 3 m.

    155 kPa; 128 kPa

12.11 The suction line of a pumping system pumping sea water (*RD* 1.04) at a rate of 80 L/s is a pipe of diameter 200 mm. A pressure gauge in this line, 2 m below the centreline of the pump, reads –15 kPa. The discharge line from the pump is a pipe of diameter 150 mm and a pressure gauge, 3 m above the centreline of the pump, reads 120 kPa. Determine the power input to the pump if its efficiency is 60%,

(a) neglecting losses;

(b) if the head loss between the two gauges is 5m.

    (a) 25.8 kW   (b) 32.6 kW

12.12 A water turbine/pump (efficiency 94%) is coupled to an electric generator/motor (efficiency 95%) and installed between a reservoir and a lake as shown in Figure P12.12. Head losses are 1.6 m when the flow velocity is 3.6 m/s. Determine:

(a) the electrical power output when the water is flowing from the reservoir to the lake;

(b) the electrical power input when the water is flowing from the lake to the reservoir.

    (a) 478 kW   (b) 742 kW

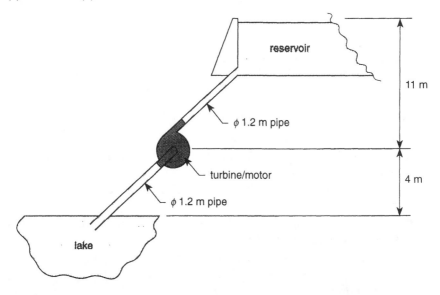

Fig. P12.12

12.13 Water is pumped at a rate of 60 L/s through the system illustrated in Figure P12.13. Head loss in the suction line is 0.8 m and in the delivery line is 4.5 m. The pump has an efficiency of 72%. Determine:

(a) the pressure at the inlet to the pump (p_1);

(b) the pressure at the outlet to the pump (p_2);

(c) the power required to drive the pump.

    (a) 9.95 kPa   (b) 142 kPa   (c) 11.75 kW

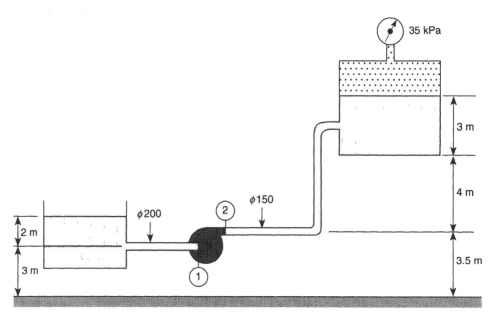

Fig. P12.13

12.14 A fan is used to draw air (specific volume 0.87) from outside a building through a ducting system and deliver it through grilles in rooms as shown in Figure P12.14. A water manometer just ahead of the fan at ② reads –80 mm and at the outlet to the fan ③ reads 140 mm. Specific-volume and potential-energy changes of the air are negligible. Determine:

(a) the pressure at ② and ③ in Pa;

(b) the head loss between ① and ② (apply Bernoulli's equation between these positions);

(c) the head loss between ③ and ④ (apply Bernoulli's equation between these positions);

(d) the increase in head across the fan using the pressures calculated in part (a);

(e) the air power delivered by the fan;

(f) the input power needed to drive the fan if the efficiency of the fan is 75%.

    (a) –785 Pa; 1373 Pa   (b) 64.5 m   (c) 115.4 m   (d) 191.4 m   (e) 5.4 kW
    (f) 7.19 kW

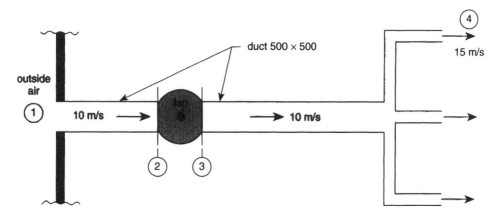

Fig. P12.14

12.15 Solve part (d) of Problem 12.14 by applying Bernoulli's equation between ① and ④. Also determine the daily energy cost of operating the fan for 12 hours per day if the electric motor driving the fan is 85% efficient and electricity costs 11 cents per kWh.
191.4 m; $11.17

Forces developed by flowing fluids

Objectives

On completion of this chapter you should be able to:

- write the impulse–momentum equation for the force associated with a change in momentum of a flowing fluid and calculate this force in typical engineering applications;
- use the impulse–momentum equation to calculate the force developed by a fluid jet striking a surface that may be perpendicular, inclined, curved, stationary or moving;
- calculate the force, torque and fluid power developed when a jet strikes a series of moving blades (impulse turbine), and calculate the efficiency;
- calculate the forces developed by a fluid flowing in a duct or pipe with changes in cross-section and/or direction.

Introduction

In Chapter 10 the forces associated with a static fluid were treated. These forces were due to static pressure and/or self-weight. When a fluid flows, additional forces occur. These forces arise from a basic Newtonian law of mechanics, which states that a body will move in a straight line with constant velocity unless acted on by an external force. This law applies to fluids as well as solids and therefore whenever the velocity of a fluid changes in either magnitude or direction, a force is involved. It is also a basic law of mechanics that forces occur in pairs, that is, every action force has an equal and opposite reaction force. This means that the force acting *on* a fluid to change its velocity results in an equal and opposite force exerted *by* the fluid. For example, a propeller exerts a force on a fluid to change its velocity, and the fluid exerts an equal and opposite force on the propeller. The reactive force on the propeller provides the propulsive force.

In this chapter, the forces developed by a free jet of fluid are treated first. Then the forces developed by an enclosed fluid are considered.

13.1 *THE IMPULSE–MOMENTUM EQUATION*

For any body with mass, the force associated with a change in velocity can be determined from the relationship:

Impulse = change in momentum

$$\therefore Ft = m(v_2 - v_1)$$

or

$$F = \frac{m}{t}(v_2 - v_1)$$

For a fluid $\dfrac{m}{t}$ = mass flow rate $\dot{m}$

Therefore, for a fluid, the force associated with a change in velocity is given by:

$$\boxed{F = \dot{m}\,(v_2 - v_1)}$$ fluid dynamics force (13.1)

Where: F = force *on* the fluid in N
$\dot{m}$ = mass-flow rate of the fluid in kg/s
v_2 = final velocity of the fluid in m/s
v_1 = initial velocity of the fluid in m/s

Notes

1 Equation 13.1 must be applied in a *certain direction*, which should be specified. usually, two mutually perpendicular directions (*x* and *y* directions) are used. The resultant force may then be calculated using trigonometry (Pythagoras's theorem).

2 The force, as given by Equation 13.1, is the force *on* the fluid. The reaction force exerted *by* the fluid will be equal and opposite.

This equation will now be used in a variety of situations to calculate fluid dynamic forces.

Example 13.1

The propeller of an aircraft is 2 m in diameter. The aircraft is flying at a speed of 250 km/h at an altitude where the air density is 0.65 kg/m³. If the velocity of the air leaving the propeller (relative to the aircraft) is 500 km/h, determine the thrust developed by the propeller. Assume that the density of the air is constant, and that the velocity of the air at the propeller itself is the mean of the upstream and downstream velocities. Neglect frictional losses.

Solution

Relative to the aircraft the velocities of the air are as shown in Figure 13.1.

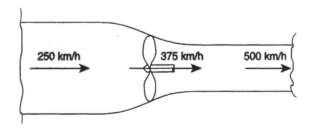

Fig. 13.1

The mass-flow rate of air through the propeller may be calculated from the average (or mean velocity) at the propeller itself and velocities in km/h may be converted to m/s by dividing by 3.6.

Now $\dot{m} = vA\rho = \dfrac{375}{3.6} \times A(2) \times 0.65 = 212.7$ kg/s

The force on the air is given by $F = \dot{m}(v_2 - v_1)$

$$= 212.7 \times \frac{500 - 250}{3.6} \text{ N}$$

$$= 14.8 \text{ kN } (+ \text{ therefore } \rightarrow)$$

Therefore, the thrust on the propeller is **14.8 kN** $\leftarrow$

Self-test problem 13.1

In a rocket of mass 1.2 t, the combustion gases exit from the rocket at atmospheric pressure and with a velocity relative to the rocket of 560 m/s. The diameter of the gas stream is 350 mm and the density of the gas is 0.55 kg/m³. Determine:
(a) the thrust produced by the rocket engine;
(b) the upward acceleration of the rocket at lift-off.

13.2 *FLUID JET STRIKING A PERPENDICULAR FLAT SURFACE*

A common application of the impulse–momentum equation involves a fluid jet striking a surface and being deflected by it. The case of a flat surface perpendicular to the jet stream will now be treated.

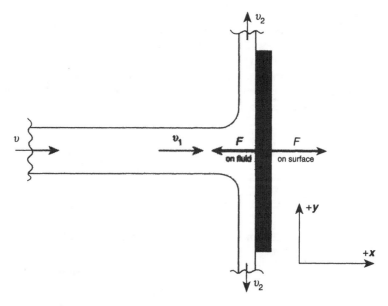

Fig. 13.2   *Jet striking a perpendicular flat surface*

Assume the jet stream is moving horizontally with velocity v and when this stream strikes a flat surface perpendicular to it, the flow is as shown in Figure 13.2. Conventional x and y directions are shown on this diagram and the intial and final velocities in these directions are as follows:

x direction Initial velocity $v_1 = v$ (the velocity of the jet stream)

Final velocity $v_2 = 0$ (the jet stream is moving upward and there is no x component of the velocity)

y direction Initial velocity $v_1 = 0$ (no y component of the velocity)

Final velocity $v_2 = 0$ (the upward and downward motions cancel out so there is no nett velocity in the y direction)

Therefore the force in the y direction is zero.

Example 13.2

Water discharges horizontally at a rate of 5 kg/s through a nozzle. The jet, which is 25 mm in diameter, strikes a vertical flat plate. Determine the force on the plate (in magnitude and direction).

Solution

$$\dot{m} = 5 \text{ kg/s}$$

$$v = \frac{\dot{m}}{\rho A} = \frac{5}{1000 \times A(0.025)} = 10.186 \text{ m/s}$$

In the x direction:

$$v_1 = v = 10.186 \text{ m/s}, \ v_2 = 0$$
$$F = \dot{m}(v_2 - v_1)$$
$$\therefore F = 5 \times (0 - 10.186)$$
$$= -50.9 \text{ N}$$

The negative sign indicates that the force on the fluid acts to the left. Therefore the force on the plate is positive (acts to the right).

Hence the force on the plate is **50.9 N**

In the y direction:

There is no force in the y direction as proved in Section 13.2 above.

Self-test problem 13.2

Compressed air leaves a nozzle of diameter 5 mm with velocity 50 m/s and strikes a flat surface perpendicular to it. Determine the force on the surface if the density of the air is 1.2 kg/m³.

13.3 FLUID JET STRIKING AN INCLINED FLAT SURFACE

A jet of fluid striking an inclined flat surface is illustrated in Figure 13.3 on page 270.

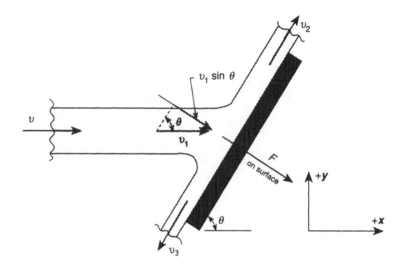

Fig. 13.3   *Jet striking an inclined flat surface*

Now the jet stream will not divide equally and more fluid will flow up the surface then down it. However, if friction forces are ignored, there is no force parallel to the surface. Hence the total force (or resultant force) on the surface is perpendicular to it. If the horizontal and vertical forces are needed, this force may be resolved into its x and y components.

As the impulse–momentum equation may be applied in any direction it is convenient in this case to use a direction perpendicular to the surface.

Perpendicular direction   Initial velocity = $v_1 \sin \theta$ (see Figure 13.3)
Final velocity = 0

Example 13.3

A horizontal jet of water of diameter 10 mm and moving with velocity 15 m/s strikes a plate inclined upward at an angle of 60° (refer to Figure 13.3). Determine the force on the plate and state the direction of the force relative to the x (horizontal) axis.

Solution

$$\dot{m} = vA\rho = 15 \times A(0.010) \times 10^3 = 1.178 \text{ kg/s}$$

In the normal (or perpendicular) direction:

$$v_1 = v \sin \theta = 15 \sin 60° = 13 \text{ m/s}$$
$$F = \dot{m}(v_2 - v_1) = 1.178 \times (0 - 13) = -15.3 \text{ N (on fluid)}$$

Because there is no force parallel to the plate, the resultant force on the plate is **15.3 N** at an angle of –30°.

Self-test problem 13.3

A house roof is inclined at an angle of 26° to the horizontal. A fire hose discharges a jet of water 35 mm in diameter horizontally at 45 km/h toward one side of the roof.

Determine the total force and also the horizontal and vertical forces on the roof surface caused by the impact of the jet. Neglect friction.

13.4 *FLUID JET STRIKING A CURVED SURFACE*

When a surface is curved, the method of calculating the force follows the same general method as for an inclined surface. This is because the resultant force due to a change in momentum of a jet of fluid depends only on the initial and final conditions and the intermediate curvature does not affect it. However, if the surface is curved so that the jet moves backward after impact, care is needed to assign the correct sign to the velocity. For example, consider a jet that is deflected through 180°. If the jet was initially moving horizontally from left to right, $v_1 = v$ but, after deflection, the jet is moving horizontally from right to left so $v_2 = -v$. The force is then given by:

$$F = \dot{m}(v_2 - v_1) = \dot{m}(-v - v) = -2mv$$

which is *twice* the force caused by a deflection of 90° (see Section 13.2).

Example 13.4

A steam jet 35 mm in diameter is discharged from a nozzle with velocity 80 m/s and specific volume 2.3 m³/kg. The jet strikes a stationary turbine blade, which deflects the steam as shown in Figure 13.4. Calculate the magnitude and direction of the resultant force on the blade, if friction is neglected.

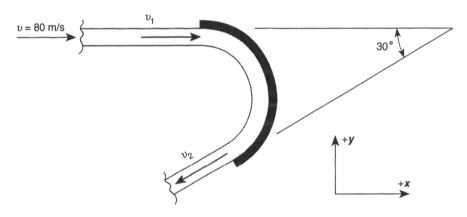

Fig. 13.4

Solution

The mass-flow rate is:

$$\dot{m} = vA\rho = 80 \times A(0.035) \times \frac{1}{2.3} = 0.0335 \text{ kg/s}$$

In the x direction:
Initial velocity $v_1 = v = 80$ m/s
Final velocity $v_2 = -80 \cos 30° = -69.3$ m/s (negative because it acts backward)

The force on the fluid in the x direction is:

$$Fx = \dot{m}(v_2 - v_1) = 0.0335 \times (-69.3 - 80) = -5 \text{ N} \leftarrow$$

In the y direction:
Initial velocity $v_1 = 0$
Final velocity $v_2 = -80 \sin 30° = -40$ m/s (negative because it acts downward)

The force on the fluid in the y direction is:

$$Fy = \dot{m}(v_2 - v_1) = 0.0335 \times (-40 - 0) = -1.34 \text{ N} \downarrow$$

Resultant force:

The x and y forces on the fluid may now be combined as shown in Figure 13.5.

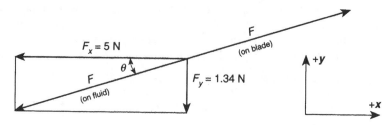

Fig. 13.5

Angle θ is given by:

$$\tan \theta = \frac{1.34}{5}$$

$$\therefore \theta = 15°$$

and $\dfrac{F_x}{F} = \cos \theta \therefore F = \dfrac{F_x}{\cos \theta} = \dfrac{5}{\cos 15°} = \textbf{5.18 N}$

Hence the force on the blade has a magnitude of **5.18 N** and acts upward to the right at an angle of **15°** to the horizontal.

Note This angle is as would be expected because it is half the deflection angle of 30° (refer to Fig. 13.3).

 ## Self-test problem 13.4

A vertical jet of water of diameter 12 mm and with velocity 20 m/s strikes a curved cup as shown in Figure 13.6. Determine the force on the cup if:
(a) frictional losses are neglected;
(b) friction causes a reduction in velocity of 5% in the water leaving the cup.

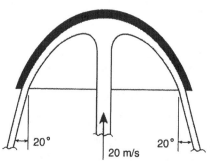

Fig. 13.6

13.5 *FLUID JET STRIKING A MOVING SURFACE*

When the surface is moving relative to the jet, the method is the same except that it is necessary to use mass-flow rates and velocities *relative* to the surface.

Note If the surface is moving *away* from the jet, the relative velocity is obtained by *subtraction* and if the surface is moving *toward* the jet, the relative velocity is obtained by *addition*.

Example 13.5

For the turbine blade given in Example 13.4 recalculate the force on the blade in the x and y directions if the turbine blade is moving horizontally:
(a) toward the right with velocity 20 m/s;
(b) toward the left with velocity 20 m/s.

Solution

(a) The blade is moving away from the jet so the relative velocity is: $80 - 20 = 60$ m/s
   The relative mass-flow rate is:

$$\dot{m} = vA\rho = 60 \times A(0.035) \times \frac{1}{2.3} = 0.0251 \text{ kg/s}$$

In the x direction:
Initial relative velocity $v_1 = 60$ m/s
Final relative velocity $v_2 = -60 \cos 30° = -51.96$ m/s

The force on the fluid in the x direction is:

$$Fx = \dot{m}(v_2 - v_1) = 0.0251 \times (-51.96 - 60) = -2.81\text{N} \leftarrow$$
$$\therefore \text{ the force on the blade in the } x \text{ direction is } \mathbf{2.81 \ N} \rightarrow$$

In the y direction:
Initial relative velocity $v_1 = 0$
Final relative velocity $v_2 = -60 \sin 30° = -30$ m/s $\downarrow$

The force on the fluid in the y direction is:

$$Fy = \dot{m}(v_2 - v_1) = 0.0251 \times (-30 - 0) = -0.753 \text{ N} \downarrow$$
$$\therefore \text{ the force on the blade in the } y \text{ direction is } \mathbf{0.753 \ N} \uparrow$$

(b) The blade is moving toward the jet so the relative velocity is: $80 + 20 = 100$ m/s
   The relative mass-flow rate is:

$$\dot{m} = vA\rho = 100 \times A(0.035) \times \frac{1}{2.3} = 0.0418 \text{ kg/s}$$

In the x direction:
Initial relative velocity $v_1 = 100$ m/s
Final relative velocity $v_2 = -100 \cos 30° = -86.6$ m/s

The force on the fluid in the x direction is:

$$Fx = \dot{m}(v_2 - v_1) = 0.0418 \times (-86.6 - 100) = -7.806 \text{ N} \leftarrow$$
$$\therefore \text{ the force on the blade in the } x \text{ direction is } \mathbf{7.81 \ N} \rightarrow$$

In the y direction:
Initial relative velocity $v_1 = 0$

Final relative velocity $v_2 = -100 \sin 30° = -50$ m/s ↓

The force on the fluid in the y direction is:

$$Fy = \dot{m}(v_2 - v_1) = 0.0418 \times (-50 - 0) = -2.09 \text{ N} \downarrow$$
$\therefore$ the force on the blade in the y direction is **2.09 N** ↑

 Self-test problem 13.5

Repeat Self-test problem 13.4 if the cup is moving upward with velocity 5 m/s.

13.6 *FLUID JET STRIKING A SERIES OF MOVING SURFACES*

In a turbine, a series of buckets or blades are mounted around a disc or wheel in such a way that they continuously intercept and redirect the velocity of the fluid jet. This means that a virtually continuous force is obtained at the pitch diameter of the buckets or blades and this force provides the torque and power on the output shaft.

If the blades are shaped so that only a change in direction of the velocity occurs, the turbine is known as an impulse turbine. The general arrangement of such a turbine is shown in Figure 13.7.

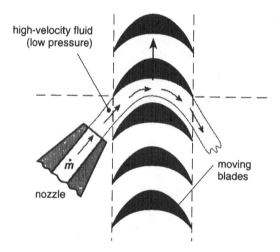

Fig. 13.7   *Impulse turbine*

An impulse turbine, used with water, that has shaped blades (or buckets) is also known as a Pelton wheel.

Many turbines used with gases have blades shaped so that the fluid accelerates while passing through the blades. As mentioned in Chapter 6, these turbines are known as impulse–reaction turbines. The acceleration of the fluid in the moving blades (increase in velocity) is accompanied by a reduction in pressure (as given by Bernoulli's equation) unlike impulse turbines, in which no such pressure reduction occurs once the fluid leaves the nozzle.

The calculations involved with impulse–reaction turbines is beyond the scope of this book, but impulse turbines may be treated using the principles and methods developed thus far. The only difference is that where there is a series of moving blades and the fluid leaving the jet is being intercepted continuously, the absolute mass-flow rate is used in

Equation 13.1 and not the relative mass-flow rate. However, velocities relative to the blades are still used in this equation. This is illustrated in Example 13.6.

Example 13.6

The nozzle of a Pelton-wheel turbine discharges a water jet of diameter 35 mm with velocity 50 m/s. The buckets are shaped as shown in Figure 13.8 and are moving with a velocity of 25 m/s. Determine the output power and efficiency (overall). Assume that fluid friction causes a 5% head loss to the turbine, a 5% reduction in velocity of flow over the buckets and that mechanical friction losses are 5%.

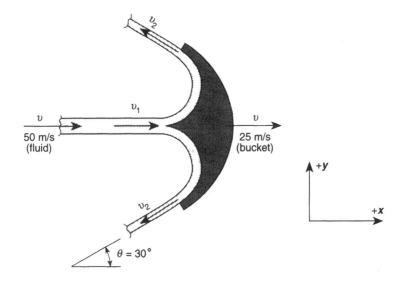

Fig. 13.8

Solution

The absolute mass-flow rate is: $\dot{m} = vA\rho = 50 \times A(0.035) \times 1000 = 48.1$ kg/s
The x direction only needs to be considered because the blades are shaped so that there is no force in the y direction (this eliminates axial loads on the shaft). In the x direction:

The relative velocities are: $v_1 = 50 - 25 = 25$ m/s

$$v_2 = -0.95 \, v_1 \cos 30° = -0.95 \times 25 \times \cos 30° = -20.57 \text{ m/s}$$
$$F = \dot{m}(v_2 - v_1) = 48.1 \times (-20.57 - 25) \text{ (N)} = -2.192 \text{ kN (on fluid } \leftarrow\text{)}$$
$$\therefore \text{ force on the blade is 2.192 kN} \rightarrow$$

From Equation 3.8, $P = Fv$

$$\therefore \text{ Power developed at the blades is } 2.192 \times 25 = 54.8 \text{ kW}$$

Because mechanical friction loses are 5%,

$$\text{Output power} = 0.95 \times 54.8 = \textbf{52.06 kW}$$

The fluid power of the water leaving the nozzle is due to the velocity head (because the stream is at atmospheric pressure).

$$\text{Now } h_v = \frac{v^2}{2g} = \frac{50^2}{2 \times 9.81} = 127.42 \text{ m}$$

Because the head loss is 5%, the initial head of the water is $\frac{127.42}{0.95} = 134.1$ m

Initial fluid power $P = \dot{m}gH = 48.1 \times 9.81 \times 134.1$ (W) $= 63.3$ kW

Therefore the overall efficiency of the turbine is $\frac{52.06}{63.3} = \mathbf{82.25\%}$

13.7 *ENCLOSED FLUIDS*

When the fluid is enclosed (passing through a duct or pipe), the analysis of dynamic fluid forces is more complex due to the fact that pressure forces now also act. In the case of a free-fluid jet, the fluid was at atmospheric pressure so that static-fluid-pressure forces did not need to be considered. However, if the fluid is passing through a duct or pipe of changing section, there is a change in pressure associated with the change in velocity and this needs to be taken into account. For example, consider the fluid pressure forces that occur when a fluid flows through a nozzle as shown in Figure 13.9.

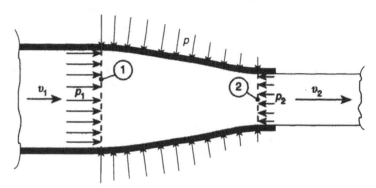

Fig. 13.9

Consider the mass of fluid in the nozzle (at any given instant) between the upstream location ① and the downstream location ② just at the exit to the nozzle. There is a pressure force p_1A_1 acting to the right ($+x$ direction) and a pressure force p_2A_2 acting to the left ($-x$ direction). Also there is a pressure force acting along the tapered section and the x component of this force is the resultant force acting on the nozzle. The pressure varies between locations ① and ② but the force on the nozzle may be calculated by use of the impulse–momentum equation.

Let the resultant force on the fluid be F (assumed positive).

Then the impulse–momentum equation applied in the x direction yields:

$$p_1A_1 - p_2A_2 + F = \dot{m}(v_2 - v_1)$$

This equation enables the resultant force on the fluid element in the x direction to be calculated and the force on the nozzle will be equal and opposite. This is illustrated in Example 13.7.

Example 13.7

For the nozzle shown in Figure 13.9, pressure p_1 is 800 kPa (gauge) and the diameter is 45 mm. Pressure p_2 is 50 kPa (gauge) and the diameter is 30 mm. Water flows through the nozzle at a rate of 30 L/s. Determine the horizontal force on the nozzle.

Solution

Velocities are: $v_1 = \dfrac{0.03}{A(0.045)} = 18.86$ m/s $\qquad v_2 = \dfrac{0.03}{A(0.03)} = 42.44$ m/s

Substituting in the impulse–momentum equation:

$$p_1 A_1 - p_2 A_2 + F = \dot{m}(v_2 - v_1)$$
$$\therefore\ 800 \times 10^3 \times A(0.045) - 50 \times 10^3 \times A(0.03) + F = 30 \times (42.44 - 18.86)$$
$$\therefore\ 1272.3 - 35.3 + F = 707.3$$
$$\therefore\ F = -530 \text{ N} \leftarrow \text{(on fluid)}$$
$$\therefore\ F = \mathbf{530\ N} \rightarrow \text{(on nozzle)}$$

Self-test problem 13.6

Water flows through a horizontal transition piece as shown in Figure 13.10. Downstream of the transition the pressure is 200 kPa (gauge). If frictional losses are neglected, determine:

(a) the upstream pressure;

(b) the horizontal force on the transition piece (stating direction).

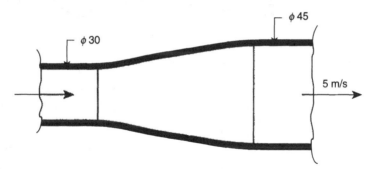

Fig. 13.10

Example 13.8

Calculate the forces in the x and y directions if the nozzle in Example 13.7 is bent upward at an angle of 30° at exit.

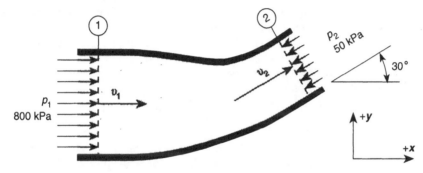

Fig. 13.11

Solution

In the x direction the impulse–momentum equation is:

$$p_1A_1 - p_2A_2 \cos 30° + F = \dot{m}(v_2 \cos 30° - v_1)$$

$$\therefore 800 \times 10^3 \times A(0.045) - 50 \times 10^3 \times A(0.03) \times \cos 30° + F =$$

$$30 \times (42.44 \times \cos 30° - 18.86)$$

$$\therefore 1272.3 - 30.6 + F = 536.8$$

$$\therefore F = -705 \text{ N } (\leftarrow \text{ on fluid})$$

$$\therefore F = \mathbf{705 \text{ N }} (\rightarrow \text{ on nozzle})$$

In the y direction the impulse–momentum equation is:

$$0 - p_2A_2 \sin 30° + F = \dot{m}(v_2 \sin 30° - 0)$$

$$\therefore -50 \times 10^3 \times A(0.03) \times \sin 30° + F = 30 \times (42.44 \times \sin 30°)$$

$$\therefore -17.7 + F = 636.6$$

$$\therefore F = 654 \text{ N } (\uparrow \text{ on fluid})$$

$$\therefore F = \mathbf{654 \text{ N }} (\downarrow \text{ on nozzle})$$

Summary

The force due to a flowing fluid (fluid-dynamics force) may be calculated by use of the impulse–momentum equation $F = \dot{m}(v_2 - v_1)$. This equation can be applied only in a single direction, that is, the force is in the same direction as the change in velocity. When v_2 is the final velocity and v_1 is the initial velocity, this equation gives the force *on* the fluid. The reaction force exerted *by* the fluid is equal and opposite.

When a fluid jet strikes a surface, the change in direction of the fluid results in a force being exerted by the fluid on the surface. Usually, fluid friction forces are negligible compared with momentum forces, and when this is the case, the force on the surface acts perpendicular to the surface at any point. Therefore, for a flat surface (when friction is negligible), the resultant force acts perpendicular to the surface regardless of the angle of inclination to the jet. If the surface is curved and the jet enters smoothly (parallel to the initial inclination of the surface), then the resultant force acts midway between the entry and exit angles of the jet. If the surface deflects the fluid through 180° the force is twice as large as the force that occurs when the deflection angle is 90°.

If the surface is moving, the force should be calculated using the relative mass-flow rate and the relative change in velocity. If there is a series of moving surfaces (as is the case with an impulse turbine) the force is calculated using the absolute mass-flow rate and the relative change in velocity.

When the fluid is enclosed, there may be a change in pressure as well as a change in velocity. The pressure forces acting on an element of fluid (in any direction) need to be taken into account when writing the impulse–momentum equation. The general form of the equation in this case is:

$$p_1A_1 - p_2A_2 + F = \dot{m}(v_2 - v_1)$$

If the subscripts 1 and 2 refer to initial and final conditions respectively, then this equation gives the force *on* the fluid, the force exerted *by* the fluid is equal and opposite.

Problems

For these problems neglect losses and forces due to the self-weight of the fluid. Assume that the fluid is initially moving in the positive x direction unless specified otherwise.

13.1 Water discharges at a rate of 8 kg/s through a nozzle. The jet is 30 mm in diameter and strikes a vertical wall. Determine the horizontal force on the wall.

90.5 N

13.2 A jet of water 20 mm in diameter travels vertically upward with velocity of 25 m/s when it strikes a hemispherical cup (which deflects the water through 180°). Neglecting friction, determine the force on the cup.

393 N

13.3 If friction causes a 10% reduction in the water velocity leaving the cup given in Problem 13.2, determine what the force on the cup will now be.

373 N

13.4 A jet engine in a stationary aircraft takes in 200 kg of air per second and 10 kg of fuel per second. The exhaust gas exits with a velocity of 450 m/s. Determine the thrust produced by the engine.

94.5 kN

13.5 An aircraft propeller is 1600 mm in diameter. As the aircraft moves along the runway at 60 km/h, the velocity of the air leaving the propeller is 150 km/h (relative to the aircraft). Determine the thrust developed by the propeller.

Neglect losses and assume that the specific volume of the air is 0.85 m³/kg and is constant and that the velocity of the air at the propeller itself is the mean of the upstream and downstream values.

1.72 kN

13.6 A jet ski boat draws in sea water (*RD* 1.05) and discharges it with a velocity of 30 km/h relative to the boat as shown in Figure P13.6. Determine:

(a) the forward thrust if losses are neglected;

(b) the power needed to drive the propeller if fluid friction and mechanical friction losses are 20%. Neglect the potential heat change.

(a) 573 N   (b) 2.98 kW

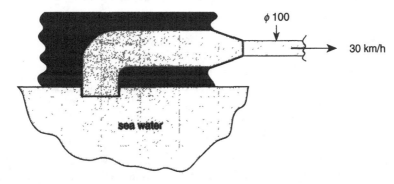

Fig. P13.6

13.7 A jet of water of cross-section 50 mm × 10 mm is deflected vertically upward through an angle of 90° by a fixed curved vane. The water velocity is 20 m/s and may be assumed to be constant. Determine the horizontal force on the vane and also the magnitude and direction of the resultant force on the vane.

> 200 N; 283 N at 45° downward

13.8 A jet of oil 50 mm in diameter travelling with velocity 30 m/s strikes a flat plate whose perpendicular axis is inclined clockwise at 30° to the jet. If the oil has a relative density of 0.92, determine:
(a) the normal force on the plate;
(b) the force on the plate in directions parallel to and perpendicular to the jet.

> (a) 1.41 kN   (b) 1.22 kN; −704 N

13.9 A jet of water 80 mm in diameter travelling with velocity 20 m/s strikes a curved vane, which deflects it downward through an angle of 120°. Determine:
(a) the horizontal force on the vane;
(b) the vertical force on the vane;
(c) the resultant force on the vane (magnitude and direction).

> (a) 3.02 kN   (b) 1.74 kN   (c) 3.48 kN at 30° upward

13.10 Suppose in Problem 13.9, a series of vanes (rather than a single vane) was mounted on a wheel of pitch diameter 1800 mm, rotating about a horizontal axis with peripheral velocity 10 m/s. Determine:
(a) the torque;
(b) the power;
(c) the efficiency.

> (a) 1357 Nm   (b) 15.08 kW   (c) 75%

13.11 A jet of water 50 mm in diameter travelling at 18 m/s strikes a flat plate held perpendicular to it. Determine the force on the plate when:
(a) the plate is stationary;
(b) the plate moves with velocity 6 m/s away from the jet;
(c) the plate moves with velocity 6 m/s towards the jet.

> (a) 636 N   (b) 283 N   (c) 1.13 kN

13.12 If the plate given in problem 13.8 moves horizontally with velocity 10 m/s away from the jet, determine:
(a) the normal force on the plate;
(b) the work done per second (power) of the moving plate.

> (a) 626 N   (b) 5.42 kW

13.13 A jet of water 50 mm in diameter moving at 18 m/s strikes a plate inclined anti-clockwise 115° from the horizontal (25° from the vertical). Determine:
(a) the normal (perpendicular) force on the plate when it is stationary;
(b) the normal (perpendicular) force on the plate when it is moving horizontally with velocity 4.5 m/s away from the jet;
(c) the power developed at the plate in part (b).

> (a) 577 N   (b) 324 N   (c) 1.32 kW

13.14 The buckets of a Pelton wheel deflect 10 kg/s of water travelling with velocity 60 m/s through an angle of 160°. Neglecting friction, determine the power developed by the wheel when the peripheral wheel velocity varies from 0 to 60 m/s in steps of 10 m/s. Plot these points on a graph and show that maximum power is developed when the wheel velocity is one-half the jet velocity. (*Note* This is always the case.)

0; 9.7 kW; 15.5 kW; 17.5 kW; 15.5 kW; 9.7 kW; 0

13.15 Oil (*RD* 0.92) flows with velocity 3 m/s along a horizontal pipe of diameter 200 mm. The upstream pressure is 200 kPa (gauge) and the head loss in the pipe is 4.5 m. Determine the horizontal force on the pipe and state how this is transmitted from the fluid to the pipe.

1.28 kN

13.16 A horizontal pipe of diameter 250 mm has a vertically upward 90° bend. Water flows at 10 m/s through the pipe and the pressure at the entrance to the bend is 50 kPa and at the exit to the bend 45 kPa. Determine the force on the bend:

(a) in a horizontal direction;
(b) in a vertical direction;
(c) the resultant (magnitude and direction).

(a) 7.36 kN (b) –7.12 kN (c) 10.2 kN at 44° downward

13.17 A horizontal nozzle has an outlet diameter of 40 mm and an inlet diameter of 50 mm. The pressure at the nozzle entrance is 550 kPa and at the nozzle exit is 50 kPa. Determine the thrust on the nozzle when discharging water with exit velocity 40 m/s.

293 N

13.18 If the nozzle given in Problem 13.17 is bent vertically downward at 45°, determine the horizontal, vertical and resultant forces on the nozzle.

901 N; 1.47 kN; 1.72 kN at 58.4° upward

13.19 The arm shown in Figure P13.19 is free to rotate about a vertical axis. Water at a pressure of 350 kPa (gauge) enters the arm vertically and leaves as a jet with diameter 15 mm. Determine the torque developed by the arm if head losses are 15%.

52.6 Nm

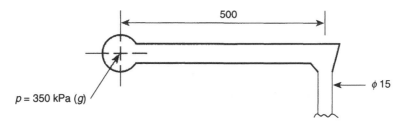

500

ϕ 15

p = 350 kPa (*g*)

Fig. P13.19

13.20 A Pelton-wheel turbine as shown in Figure P13.20 (on page 282) has buckets that deflect the water through an angle of 160°. The buckets are mounted on a wheel that rotates at 115 rpm at a pitch diameter of 1200 mm. The flow rate is adjusted to a rate of 20 L/s. Neglecting losses, determine:

(a) the velocity of the water leaving the nozzle;
(b) the force developed at the buckets;
(c) the output torque;
(d) the output power;
(e) the efficiency.

(a) 14 m/s (b) 263 N (c) 158 Nm (d) 1.9 kW (e) 96.9%

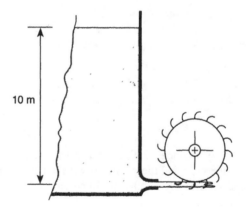

Fig. P13.20

13.21 Repeat Problem 13.20, with the same flow rate and wheel speed, if the following losses occur:

- a head loss of 5%;
- a reduction in velocity of 5% of the water flowing over the buckets (due to friction);
- a power loss of 5% due to mechanical friction (and air friction) at the wheel and shaft.
  (a) 13.65 m/s   (b) 243 N   (c) 139 Nm   (d) 1.67 kW   (e) 85.1%

Solutions to self-test problems

Chapter 1

1.1 **(a)** Efficiency of conversion is the ratio of useful energy out of the conversion process to the energy input to the process.

(b) 1 To reduce costs because the higher the conversion efficiency, the lower the cost of the input energy. Also, a smaller plant is needed, thus reducing capital costs (first costs).

2 To save energy (reduce wastage) and enable our energy reserves to last longer.

(c) Direct conversion of energy is the conversion of energy from the available form into the required form without any intermediate steps. Advantages are higher efficiency with fewer conversion steps and less equipment is usually needed.

(d) No, 100% efficiency of conversion is not possible with heat engines due to the limitations of the second law of thermodynamics.

(e) Parallel: $\eta = 0.8$ or 80%; in series: $\eta = 0.8 \times 0.8 \times 0.8 = 0.512$ or 51.2%.

1.2 **(a)** Disadvantages are: difficult to control combustion pollutants, cannot be used in internal combustion engines, time taken to stabilise combustion and temperatures.

(b) Coal can be converted into a liquid fuel by the fuel–oil conversion process; coal can be converted into a gas by heating the coal in a reduced atmosphere (insufficient oxygen) either in an industrial plant or using underground gasification.

(c) Motor cars (urban) C or E; rail C, E or F; boats (small) B, C or E; ships (large) D, F and possibly B (wind assistance); aircraft (large) E.

Chapter 2

2.1 **(a)** $p = \rho g h = 7800 \times 9.81 \times 0.11$ (Pa) = **8.42 kPa (g)**

Alternatively:

volume of piston = $A(0.12) \times 0.11 = 1.244 \times 10^{-3}$ m^3

mass of piston = $\rho V = 7800 \times 1.244 \times 10^{-3} = 9.704$ kg

$$\therefore \text{ pressure} = \frac{W_t}{A} = \frac{9.704 \times 9.81}{A(0.12)} = \textbf{8.42 kPa}$$

(b) $P_{(abs)} = 8.42 + 101.3 = \textbf{109.7 kPa}$

(c) pressure due to 5 kg mass $= \dfrac{F}{A} = \dfrac{5 \times 9.81}{A(0.12)}$ (Pa) $= 4.34$ kPa

$\therefore P_{(g)} = 8.42 + 4.34 = \mathbf{12.76 \text{ kPa}}$
$P_{(abs)} = 109.7 + 4.34 = \mathbf{114 \text{ kPa}}$

2.2 **(a)** $-269 + 273 = \mathbf{4 \text{ K}}$, $15 + 273 = \mathbf{288 \text{ K}}$, $224 + 273 = \mathbf{497 \text{ K}}$
(b) $15 - 273 = \mathbf{-258°C}$, $226 - 273 = \mathbf{-47°C}$, $527 - 273 = \mathbf{254°C}$
(c) $(-25 - 15) = \mathbf{-40°C}$, $(35 - (-5)) = \mathbf{40°C}$, $(-20 - (-80)) = \mathbf{60°C}$

2.3 **(a)** capacity $= A(0.09) \times 0.1 \times 6 \times 1000 = \mathbf{3.817 \text{ L}}$

(b) $V_s/\text{cyl} = \dfrac{3.817}{6} = 0.6362 \text{ L}$

now $\dfrac{V_s + V_c}{V_c} = 9.5$, $\therefore V_s + V_c = 9.5 \, V_c$, $\therefore V_s = 8.5 \, V_c$

$\therefore 0.6362 = 8.5 \, V_c$, $\therefore V_c = \mathbf{0.0748 \text{ L}}$

(c) $5000 \text{ rpm} = \dfrac{5000}{60} \text{ rps}$

1 revolution = 2 strokes

$\therefore \text{strokes/sec} = \dfrac{5000}{60} \times 2 = \mathbf{166.\dot{6}}$

Chapter 3

3.1 **(a)** $W = Fx = 5000 \times 9.81 \times 20$ (J) $= \mathbf{981 \text{ kJ}}$
(b) $PE = mgh = 5000 \times 9.81 \times 20$ (J) $= \mathbf{981 \text{ kJ}}$
(c) Dropping 5 m, $PE = 5000 \times 9.81 \times 5$ (J) $= 245.25$ kJ

$PE = KE = \frac{1}{2}mv^2$

$\therefore \frac{1}{2} \times 5000 \times v^2 = 245.25 \times 10^3$
$\therefore v = \mathbf{9.9 \text{ m/s}}$

3.2 The work done is given by the crosshatched area in Figure S1.
Area$= p_2(V_2 - V_1) + 0.5(p_1 - p_2)(V_2 - V_1)$
$W = 100 \times 10^3 \times 0.5 + 0.5 \times (200 \times 10^3) \times 0.5$
$= 100 \times 10^3$ J
$= \mathbf{100 \text{ kJ}}$

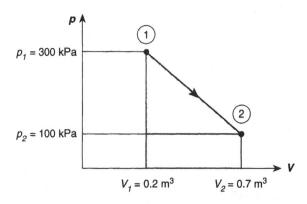

Fig. S1

3.3 (a) $P = T\omega$

$$= 3.5 \times 10^{-3} \times \frac{\pi \times 4500}{30}$$

$$= \mathbf{1.65 \ kW}$$

(b) $\eta = \dfrac{P_{out}}{P_{in}} = 0.9$

$$\therefore \ P_{in} = \frac{P_{out}}{0.9} = \frac{1.65}{0.9} = \mathbf{1.83 \ kW}$$

3.4 Ice $Q_{sensible} = mc\Delta T = 0.6 \times 2.04 \times 20 = 24.48 \ kJ$
$$Q_{latent} = mL = 0.6 \times 335 = 201 \ kJ$$
$$Q_{Total} \ ice = 225.48 \ kJ$$
$$\therefore \ Q_{To \ water} = 250 - 225.48 = 24.52 \ kJ$$
$$Q_{sensible \ water} = mc\Delta T$$
$$\therefore \ 24.52 = 0.6 \times 4.19 \times \Delta T$$
$$\therefore \ \Delta T = 9.75$$
$\therefore$ Final temperature of water = **9.75°C**

3.5 $\eta = \dfrac{P_{out}}{P_{in}} \ \therefore \ 0.33 = \dfrac{80}{P_{in}} \ \therefore \ P_{in} = 242.42 \ kW$

This is the energy supply rate from the fuel
$Q = mE \ \therefore$ per second 242.42 (kJ) $= m \times 46.5 \times 10^3$ (kJ)
$\therefore \ m = 0.0052 \ kg$ (per second)
$\therefore$ per h: $m = 0.0052 \times 3600 = 18.77 \ kg$
but 1 L = 0.74 kg

$$\therefore \ L/h = \frac{18.77}{0.74} = \mathbf{25.4 \ L}$$

3.6 $\eta = \dfrac{P_{out}}{P_{in}} \ \therefore \ P_{in} = \dfrac{800}{0.4} = 2000 \ MW$

$\therefore$ energy input/day = $2000 \times 3600 \times 24 = 172.8 \times 10^6 \ MJ$

(a) $\therefore$ coal used/day $= \dfrac{172.8 \times 10^6}{27} \ kg = \mathbf{6400} \ t$

(b) nuclear: $E = mc^2$
$$\therefore \ 172.8 \times 10^{12} \ (J) = m \times (3 \times 10^8)^2$$
$$\therefore \ m = 1.92 \times 10^{-3} \ kg = \mathbf{1.92 \ g}$$

Chapter 4

4.1 Using a time interval of 1 s:
(a) $Q = 3.5 - 0.8 = 2.7 \ kJ$
$$W = -0.5 \ kJ$$
$$Q - W = U_2 - U_1$$
$$\therefore \ 2.7 - (-0.5) = U_2 - U_1$$
$$\therefore \ U_2 - U_1 = 3.2 \ kJ$$
$$U_2 - U_1 \doteqdot mc\Delta T$$
$$\therefore \ 3.2 = 100 \times 4.19 \times \Delta T$$
$$\therefore \ \Delta T = 0.00764°C \text{ (temperature change per second)}$$

In order to reach 80°C, $\Delta T = 60°C$

$$\therefore t = \frac{60}{0.00764} = 7856 \text{ s}$$

$$\therefore t \text{ (h)} = \frac{7856}{3600} = \textbf{2.18 h}$$

(b) $U_2 - U_1 = mc\Delta T = 100 \times 4.19 \times 60 \text{ kJ}$
$$= \textbf{25.14 MJ}$$

4.2 Let the final temperature be $T°C$
Liquid: $Q = mc(T_2 - T_1) = 2 \times 1.15 \times 3.76 \times (T - 85)$
$$= 8.648T - 735.08$$
Container: $Q = mc(T_2 - T_1) = 0.8 \times 1.24 \times (T - 15)$
$$= 0.992T - 14.88$$
$\Sigma Q = 0 \therefore 8.648T - 735.08 + 0.992T - 14.88 = 0$
$\therefore 9.64T = 749.96$
$\therefore T = \textbf{77.8°C}$

4.3 Equilibrium occurs at 0°C
Let the mass of ice that melts be m kg
Ice: $Q_{\text{sensible}} = mc\Delta T = 1 \times 2.04 \times (0 - (-5)) = 10.20 \text{ kJ}$
$\quad Q_{\text{latent}} = mL = 335m$
Water: $Q_{\text{sensible}} = mc\Delta T = 2 \times 4.19 \times (0 - 18) = -150.84 \text{ kJ}$
Copper: $Q_{\text{sensible}} = mc\Delta T = 1.5 \times 0.39 \times (0 - 18) = -10.53 \text{ kJ}$
$\Sigma Q = 0$
$\therefore 10.20 + 335m - 150.84 - 10.53 = 0$
$\therefore 335m = 151.17$
$\therefore m = 0.451 \text{ kg}$
$\therefore$ mass of ice remaining $= 1 - 0.451 = \textbf{0.549 kg}$

4.4 $Q = mc\Delta T$
$$= (2.52 + 0.436) \times 4.175 \times (21.802 - 19.615)$$
$$= 26.99 \text{ kJ}$$
mass of fuel burnt $= 1.02 \text{ g} = 1.02 \times 10^{-3} \text{ kg}$

$$\therefore E = \frac{26.99}{1.02 \times 10^{-3}} \text{ kJ/kg}$$
$$= \textbf{26.46 MJ/kg}$$

4.5

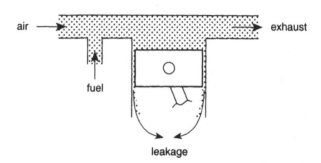

Fig. S2

Air in: $\dot{m} = \rho\dot{V} = 1.2 \times 0.625 = 0.75$ kg/s

Fuel in: $\dot{m} = \dfrac{0.75}{14.5} = 0.0517$ kg/s

$\therefore$ Total in: $\dot{m} = 0.8017$ kg/s

Exhaust gas out: $\dot{m} = \rho A v$

$$= 0.75 \times A(0.2) \times 33.7$$
$$= 0.794 \text{ kg/s}$$

$\therefore$ Leakage $\dot{m} = 0.8017 - 0.794 = \mathbf{0.0077}$ **kg/s**

4.6 (a) $\dot{Q} = \dot{m}(h_2 - h_1)$

$$= 25 \times (209 - 2838) \quad \text{kW}$$
$$= \mathbf{-65.7} \text{ MW (heat out)}$$

(b) $65.7 \times 10^3 = \dot{m}c\Delta T$

$$= \dot{m} \times 4.19 \times 24$$
$$\therefore \dot{m} = \mathbf{654} \text{ kg/s}$$

4.7 Heat transfer to the water, $Q = mc(T_2 - T_1)$

$$= 1.95 \times 4.187 \times (35.3 - 20.5)$$
$$= 120.83 \text{ kJ}$$

Gas burnt $= 6.52 - 2.85 = 3.67$ L

Energy content of the gas $= \dfrac{120.83}{3.67}$

$$= 32.9 \text{ kJ/L}$$
$$= \mathbf{32.9} \text{ MJ/m}^3$$

Chapter 5

5.1 $p_2 = 101.3$ kPa

$$= 101.3 \times 10^3 \text{ Pa}$$

$T_2 = 293$ K (same)

$V_2 = 0.1767$ m³ (same)

$R = 287$ J/kgK

$\therefore m_2 = \dfrac{p_2 V_2}{R T_2}$

$$= \dfrac{101.3 \times 10^3 \times 0.1767}{287 \times 293}$$
$$= 0.2129 \text{ kg}$$

Hence the mass of air that escapes is

$m_1 - m_2 = 2.314 - 0.2129$

$$= \mathbf{2.10} \text{ kg}$$

5.2 $m = 0.006$ kg, $p_1 = 200$ kPa, $T_1 = 100°C = 373$ K

$V_1 = 3 \times 10^{-3}$ m³, $T_2 = 300°C = 573$ K, $V_2 = 4.6 \times 10^{-3}$ m³

$c_v = 700$ J/kgK

(a) $U_2 - U_1 = mc_v(T_2 - T_1) = 0.006 \times 700 \times (300 - 100) = \mathbf{840}$ **J**

(b) $W = p(V_2 - V_1) = 200 \times 10^3 \times (4.6 \times 10^{-3} - 3 \times 10^{-3}) = \mathbf{320}$ **J**

(c) $Q - W = U_2 - U_1 \therefore Q = W + U_2 - U_1 = 320 + 840 = \mathbf{1160}$ **J**

(d) $Q = mc_p(T_2 - T_1) \therefore 1160 = 0.006 \times c_p \times 200$

$$\therefore c_p = \mathbf{966.7} \text{ J/kgK}$$

(e) Constant-pressure process: $Q = H_2 - H_1$ ∴ $H_2 - H_1 = \textbf{1160 J}$
Alternatively, $H_2 - H_1 = mc_p(T_2 - T_1) = 0.006 \times 966.7 \times 200$
$$= \textbf{1160 J}$$

5.3 $V = 35$ L $= 35 \times 10^{-3}$ m³, $T_1 = 27°C = 300$ K, $p_1 = 280$ kPa (gauge)
$$= 381.3 \text{ kPa (abs)}$$

(a) $p_1 V_1 = mRT_1$

$$\therefore m = \frac{p_1 V_1}{RT_1} = \frac{381.3 \times 10^3 \times 35 \times 10^{-3}}{287 \times 300} = \textbf{0.155 kg}$$

(b) Constant-volume process: $\dfrac{p_1}{T_1} = \dfrac{p_2}{T_2}$ ∴ $p_2 = \dfrac{T_2}{T_1} \times p_1$

$$\therefore p_2 = \frac{288}{300} \times 381.3 = 360.05 \text{ kPa (abs)} = \textbf{264.7 kPa (gauge)}$$

(c) Constant-volume process: $Q = mc_v(T_2 - T_1)$
$$= 0.155 \times 718 \times (15 - 27)$$
$$= \textbf{-1335 J}$$

(d) $U_2 - U_1 = Q$ (constant-volume process) ∴ $U_2 - U_1 = \textbf{-1335 J}$

(e) $m = \dfrac{pV}{RT} = \dfrac{381.3 \times 10^3 \times 35 \times 10^{-3}}{287 \times 288} = 0.1615$ kg

∴ mass of air to be added
$$= 0.1615 - 0.155$$
$$= \textbf{0.0065 kg}$$

5.4 (a) $\dfrac{V_2}{V_1} = \dfrac{p_1}{p_2}$ ∴ $V_2 = V_1 \times \dfrac{p_1}{p_2} = 1.5 \times \dfrac{1250}{101.3} = \textbf{18.51 L}$

(b) $W = p_1 V_1 \ln\left(\dfrac{V_2}{V_1}\right) = 1250 \times 10^3 \times 1.5 \times 10^{-3} \ln\left(\dfrac{18.51}{1.5}\right)$ J
$$= \textbf{4.71 kJ}$$

(c) $Q = W = \textbf{4.71 kJ}$

(d) $U_2 - U_1 = \textbf{0}$ for isothermal process

5.5 Data: $n = 1.2$
$p_1 = 98$ kPa
$T_1 = 45°C = 318$ K
$p_2 = 980$ kPa (gauge) $= 1081.3$ kPa (absolute)
$p_1 V_1{}^n = p_2 V_2{}^n$

$$\therefore \frac{V_1}{V_2} = \left(\frac{p_2}{p_1}\right)^{1/n}$$

$$= \left(\frac{1081.3}{98}\right)^{1/1.2}$$

$$= 7.395$$

That is, the compression ratio is **7.395:1**
The temperature at the top of the stroke (T_2) may be calculated from Equation 5.14:

$$\frac{T_2}{T_1} = \left(\frac{V_1}{V_2}\right)^{n-1}$$

$$\therefore T_2 = T_1 \left(\frac{V_1}{V_2}\right)^{n-1}$$

$$= 318 \times (7.395)^{1.2-1}$$

$$= \textbf{474.5 K (201.5°C)}$$

Note   An alternative solution is to use the general gas equation:

$$\frac{p_1 V_1}{T_1} = \frac{p_2 V_2}{T_2}$$

$$\therefore T_2 = T_1 \left(\frac{V_2}{V_1}\right) \left(\frac{p_2}{p_1}\right)$$

$$= 318 \times \left(\frac{1}{7.395}\right) \times \left(\frac{1081.3}{98}\right)$$

$$= \textbf{474.5 K}$$

The work done may be calculated from the alternative form of Equation 5.16:

$$W = \frac{mR}{n-1}(T_1 - T_2)$$

Since $m = 1$ kg,

$$W = \frac{1 \times 287}{1.2 - 1} \times (318 - 474.5) \text{ J}$$

$$= \textbf{-224.5 kJ}$$

The negative sign indicates work done on the air.

Chapter 6

6.1 (a)
$$\dot{Q}_S = \dot{Q}_R + P$$
$$= 44.8 + 17.4$$
$$= 62.2 \text{ kW}$$

$$\eta = \frac{P}{\dot{Q}_S} = \frac{17.4}{62.2} = 0.28 = \textbf{28\%}$$

(b)
$$\dot{Q}_S = \dot{m}_F E$$

$$\therefore \dot{m}_F = \frac{\dot{Q}_S}{E} = \frac{62.2}{43.1 \times 10^3} = 1.443 \times 10^{-3} \text{ kg/s}$$

$$\therefore \text{Fuel per hour} = 1.443 \times 10^{-3} \times 3600 = \textbf{5.2 kg/h}$$

6.2 $\eta_C = 1 - \dfrac{T_c}{T_h} = 1 - \dfrac{288}{T_h}$

Assuming values for T_h and substituting, the following table is obtained:

$T_h(°C)$	15	100	200	300	500	1000	1500	2000
$T_h(K)$	288	373	473	573	773	1273	1773	2273
η_C	0	0.228	0.391	0.497	0.627	0.774	0.838	0.873

Plotting these points, the graph shown in Figure S3 is obtained.

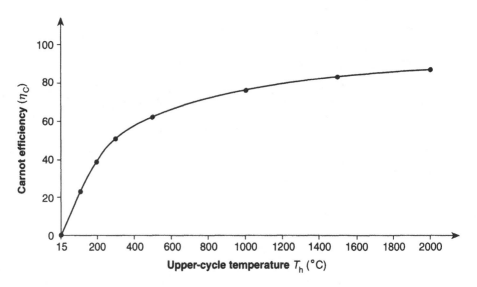

Fig. S3

It is seen that the relationship is non-linear and the improvement in efficiency between 1800°C and 2000°C is much less than the improvement in efficiency between 200°C and 400°C.

6.3	Advantages	Disadvantages
Continuous combustion	Better control over the combustion process, with less pollutants and higher combustion efficiency Simpler, more reliable and less expensive equipment needed; greater variety of fuels possible	Not suitable for reciprocating, internal-combustion engines, which require combustion to be initiated at a suitable point in each working cycle
External combustion	Better control over the combustion process with less pollutants and higher combustion efficiency Simpler, more reliable and less expensive equipment needed; greater variety of fuels possible	Combustion heat has to be transferred through the cylinder walls or by means of heat exchangers. The time taken for heat transfer limits the speed of these engines. Also, conductive materials such as metals soften with temperature and lose strength. This limits the temperature attainable by the working substance and hence the efficiency because efficiency is limited by the maximum temperature of the working substance (Carnot)

6.4 (a) $\eta_{theor} = 1 - \dfrac{T_c}{T_h} = 1 - \dfrac{293}{633} = \mathbf{0.537}$ **(53.7%)**

(b) $\dot{Q}_S = 600 \times A(1.2) = \mathbf{678.6\ W}$

(c) $P = \eta \times \dot{Q} = 0.537 \times 678.6 = \mathbf{364.5\ W}$

(d) $\dot{Q}_R = \dot{Q}_S - P = 678.6 - 364.5 = \mathbf{314.1\ W}$

6.5 First calculate a table of values:

$\dfrac{V_1}{V_2}$	1	2	3	4	5	7.5	10	15	20
$\left(\dfrac{V_1}{V_2}\right)^{\gamma-1}$	1	1.32	1.55	1.74	1.90	2.24	2.51	2.95	3.31
$T_2 = T_1 \left(\dfrac{V_1}{V_2}\right)^{\gamma-1}$ (K)	288	380	447	501	548	645	723	851	955
$\eta_{ind} = 1 - \dfrac{1}{\left(\dfrac{V_1}{V_2}\right)^{\gamma-1}}$	0	0.24	0.36	0.43	0.47	0.55	0.60	0.66	0.70

These results are plotted in Figure S4 below:

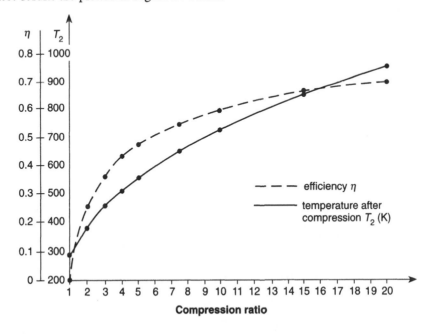

Fig. S4

Comment: It is clear that the temperature after compression and the efficiency both
increase with increasing compression ratios.

6.6 Refer to Figure 6.13.

$p_1 = 97.5$ kPa

$T_1 = 50°C = 323$ K

$$\frac{V_1}{V_2} = 20$$

$Q_s = 950$ kJ/kg

$c_p = 1.005$ kJ/kgK, $c_v = 0.718$ kJ/kgK

$\gamma = 1.4$

(a) $\dfrac{T_2}{T_1} = \left(\dfrac{V_1}{V_2}\right)^{1.4-1}$

$\therefore T_2 = 323 \times 20^{0.4}$

$= 1071$ K

For constant-pressure heating ② → ③:

$Q_s = mc_p(T_3 - T_2)$

$\therefore 950 = 1 \times 1.005 \times (T_3 - 1071)$

$\therefore T_3 = \textbf{2016 K (1743°C)}$ (maximum cycle temperature)

(b) For the constant-pressure process ② → ③:

$$\frac{V_2}{T_2} = \frac{V_3}{T_3}$$

$$\therefore V_2 = \frac{T_2}{T_3} V_3$$

$$= \frac{1071}{2016} V_3$$

$$\therefore V_2 = 0.531 V_3$$

Now,

$$\frac{V_4}{V_2} = 20$$

$$\therefore \frac{V_2}{V_4} = \frac{1}{20}$$

$$\therefore \frac{0.531 V_3}{V_4} = \frac{1}{20}$$

$$\frac{V_3}{V_4} = 0.094$$

$$\frac{T_4}{T_3} = \left(\frac{V_3}{V_4}\right)^{\gamma-1}$$

$\therefore T_4 = 2016 \times (0.094)^{0.4}$

$= 783$ K

$Q_R = mc_v(T_4 - T_1)$

$= 1 \times 0.718 \times (783 - 323)$

$= 330.5$ kJ/kg

$W = Q_S - Q_R$

$= 950 - 330.5$

$= \textbf{619.5 kJ/kg}$

(c) $\eta = \dfrac{W}{Q_S}$

$\qquad = \dfrac{619.5}{950}$

$\qquad = \textbf{65.2\%}$

Chapter 7

7.1 $T = [(8 \times 5 \times 9.81) - 9] \times \dfrac{(1.35 + 0.025)}{2}$

$\qquad = 263.6 \text{ Nm}$

$P = T\omega$

$\qquad = 263.6 \times \dfrac{\pi \times 225}{30} \text{ W}$

$\qquad = \textbf{6.21 kW}$

If one extra 5 kg mass were added, the spring-balance reading would increase to 58 N but the torque would be the same:

$T = [(9 \times 5 \times 9.81) - 58] \times \dfrac{(1.35 + 0.025)}{2}$

$\qquad = 263.6 \text{ Nm}$

If one 5 kg mass were subtracted, the spring-balance reading would tend to be negative. The applied weight would be insufficient to balance the torque output of the motor, and the weights would lift. The rope brake would now be ineffective.

7.2 $\qquad P = T\omega$

$\qquad\qquad = 200 \times 0.48 \times \pi \times \dfrac{3500}{30} \text{ W}$

$\qquad\qquad = \textbf{35.19 kW}$

$\qquad \dot{m} \text{ fuel} = \dfrac{50}{1000} \times \dfrac{0.8}{15} = 2.\dot{6} \times 10^{-3} \text{ kg/s}$

$\qquad \dot{Q}_S = \dot{m}E = 2.\dot{6} \times 10^{-3} \times 45 \times 10^3 \text{ kW}$

$\qquad\qquad = \textbf{120 kW}$

$\qquad \eta = \dfrac{P}{\dot{Q}_S} = \dfrac{35.19}{120} = \textbf{0.293 or 29.3\%}$

7.3 (a) Mass-flow rate $\dot{m}$ of fuel:

$\qquad \dot{m} = \dfrac{3.85 \times 0.9}{5 \times 60}$

$\qquad\qquad = 0.011 \, 55 \text{ kg/s}$

Heat-supply rate:

$\qquad \dot{Q}_S = \dot{m}E$

$\qquad\qquad = 0.011 \, 55 \times 42.6 \times 10^6 \text{ W}$

$\qquad\qquad = 492 \text{ kW}$

$\qquad \eta = \dfrac{P}{\dot{Q}_S}$

$\qquad\quad = \dfrac{200}{492}$

$\qquad\quad = \textbf{40.6\%}$

(b) $\text{SFC} = \dfrac{0.011\,55}{200} \times 3600$

$\qquad = \mathbf{0.208\ kg/kWh}$

7.4 $p_m = \dfrac{A_c k_s}{l_c}$

$\qquad = \dfrac{450 \times 90}{50}$

$\qquad = \mathbf{810\ kPa}$

$A = A(0.12) = 0.0113\ \text{m}^2$

$L = 0.11\ \text{m}$

$n = 6 \times \dfrac{800}{120} = 40\ \text{cycles/s}$

$IP = p_m LAn$

$\qquad = 810 \times 0.11 \times 0.0113 \times 40$

$\qquad = \mathbf{40.3\ kW}$

7.5 (a) $\eta_m = \dfrac{P}{IP}$ $\therefore P = \eta_m \times IP = 0.85 \times 40.3 = 34.255\ \text{kW}$

$\qquad\qquad\qquad\qquad\qquad\qquad\qquad$ say $\mathbf{34.3\ kW}$

(b) $FP = IP - P = 40.3 - 34.3 = \mathbf{6\ kW}$

(c) $\dot{Q}_S = \dot{m}E = \dfrac{11.3}{3600} \times 0.78 \times 46\,500 = 113.8\ \text{kW}$

$\qquad \eta = \dfrac{P}{\dot{Q}_S} = \dfrac{34.255}{113.8} = \mathbf{0.3\ (30\%)}$

(d) $\eta_{\text{ind}} = \dfrac{IP}{\dot{Q}_S} = \dfrac{40.3}{113.8} = \mathbf{0.354\ (35.4\%)}$

(e) $\text{SFC} = \dfrac{\text{kg fuel/h}}{P} = \dfrac{11.3 \times 0.78}{34.255} = \mathbf{0.257\ kg/kWh}$

7.6 (a) Swept volume/cylinder $= A(0.08) \times 0.075 \quad \text{m}^3$

$\qquad\qquad\qquad\qquad\qquad\quad = 0.377\ \text{L}$

Capacity = total swept volume

$\qquad\qquad = 4 \times 0.377$

$\qquad\qquad = \mathbf{1.51\ L}$

(b) For a four-cylinder, four-stroke:

$n = \dfrac{N}{120} \times 4$

$\quad = \dfrac{N}{30}$

where n is in cycles/s and N is in rpm.

$\therefore \dot{V}_{\text{theoretical}} = \dfrac{4000}{30} \times 0.377$

$\qquad\qquad\qquad = 50.26\ \text{L/s}$

$$\eta_V = \frac{\dot{V}_{actual}}{\dot{V}_{theoretical}}$$

$$= \frac{32}{50.26}$$

$$= \mathbf{63.7\%}$$

Chapter 8

8.1 (a) $V = Ah$

$$= A(0.56) \times 1.2$$

$$= 0.2956 \text{ m}^3$$

$$\rho = \frac{m}{V} = \frac{520}{0.2956}$$

$$\therefore \rho = \mathbf{1759 \text{ kg/m}^3}$$

$$RD = \frac{\rho}{1000} = \mathbf{1.76}$$

(b) $v = \dfrac{V}{m}$

$$\therefore m = \frac{V}{v}$$

$$\therefore m = \frac{0.2956}{0.35} = \mathbf{0.845kg}$$

8.2 $p = \dfrac{F}{A}$

$$= \frac{(1.5 + 0.5) \times 9.81}{A(0.03)} \quad \text{Pa}$$

$$= \mathbf{27.8 \text{ kPa}}$$

8.3 $\mu = \mu_{(imperial)} \times 10^{-3}$

$$= \mathbf{60 \times 10^{-3} \text{ Pas}}$$

$$v = \frac{\mu}{\rho} = \frac{60 \times 10^{-3}}{920} = \mathbf{65.2 \times 10^{-6} \text{ m}^2/\text{s}}$$

Chapter 9

9.1 (a) $A(100) = 7854 \text{ mm}^2$, perimeter $p = 314.16 \text{ mm} \therefore A/p = \mathbf{25}$
   (b) $7854 = x^2 \therefore x = 88.62 \text{ mm}$, $p = 4 \times 88.62 = 354.5 \text{ mm} \therefore A/p = \mathbf{22.16}$
   (c) $A(200) = 31\,416 \text{ mm}^2$, $p = 628.3 \text{ mm} \therefore A/p = \mathbf{50}$
   Conclusions: Round pipes are more efficient than square pipes (higher A/p)
                    Efficiency increases with size (A/p increases with size)
9.2 (a) ball valve, butterfly valve, tapered-plug cock and directional control valves
   (b) needle valve
   (c) globe valve
   (d) directional control valve
   (e) safety valve (pressure-reducing valve reduces pressure but does not necessarily fail safe)
   (f) pressure-reducing valve
   (g) foot valve or check valve

(h) diaphragm valve or butterfly valve if fitted with resilient seal

(i) as for (g)

9.3 Consider: nature of fluid (in particular corrosiveness), temperature of fluid, pressure of fluid, environment in which the gauge will be located, accuracy (over the pressure range to be measured), reliability, durability and cost.

Correct method of installation of a gauge in a pipe is shown in Figure S5.

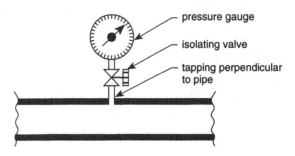

pressure gauge

isolating valve

tapping perpendicular to pipe

Fig. S5

9.4 Factors to consider: Large pipes and fittings and instruments compared with small Advantages:

• Large pipes and fittings have a lower flow loss and therefore a smaller pump can be used, which will have lower running costs (particularly energy costs).

• Expansion will be easier and less costly in the future should demand increase.

• Large instruments are generally more accurate and easier to read than small ones.

Disadvantages:

• Large pipes and fittings and instruments are more expensive to buy and install (first cost).

• Large pipes, fittings and instruments take up more space than small ones.

Chapter 10

10.1 $p = \rho g h$

$= 13.6 \times 1000 \times 9.81 \times 0.752$ Pa

$= \mathbf{100.3\ kPa}$

10.2 Vapour pressure $= \rho g h$

$= 13.6 \times 10^3 \times 9.81 \times 0.256$ Pa

$= \mathbf{34.15\ kPa}$

Maximum tank pressure is at the base of the tank

$= 34.15 \times 10^3 + 1.26 \times 10^3 \times 9.81 \times 1.3$ Pa

$= \mathbf{50.2\ kPa}$

10.3 Using $X–X$ as datum: $p_1 = p_2$, $p_2 = p_3$

$\therefore p_1 = p_3$

$p_1 = \rho g h$ (water) $= 10^3 \times 9.81 \times 0.15 \times \sin 30°$ Pa

$= 0.736$ kPa

At the centre of the pipe:

$p = p_3 + \rho g h$ (kerosene)

$= 0.736 + 0.78 \times 9.81 \times 0.2$ kPa

$= \mathbf{2.27\ kPa}$

10.4 Pressure on plate $= \rho g h$

$$\therefore p = 0.9 \times 10^3 \times 9.81 \times 2.5 \quad \text{Pa}$$
$$= 22.1 \text{ kPa}$$
$$F = pA$$
$$= 22.1 \times A(0.3) \quad \text{kN}$$
$$= \textbf{1.56 kN at the centre of the plate}$$

10.5

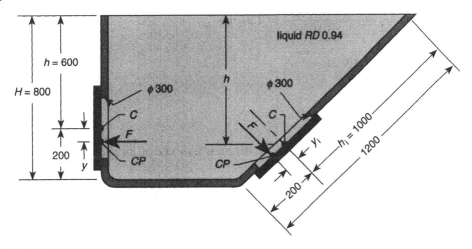

Fig. S6

(a) Vertical side: $h = 600$ mm $= 0.6$ m
$p = \rho g h = 0.94 \times 10^3 \times 9.81 \times 0.6$ Pa $= 5.53$ kPa
$F = pA = 5.53 \times A(0.3)$ kN $= \textbf{391 N}$
Using Equation 10.2

$$y = \frac{d^2}{16h} = \frac{0.3^2}{16 \times 0.6} = 0.0094 \text{ m} = 9.4 \text{ mm}$$

Therefore the force acts at a distance of $200 - 9.4 = \textbf{190.6 mm}$ vertically up from the base.

(b) For the inclined side: it is first necessary to calculate h using trigonometry.

Now $\dfrac{h}{1000} = \dfrac{800}{1200}$

Therefore $h = 667$ mm $= 0.667$ m
Now $p = \rho g h = 0.94 \times 10^3 \times 9.81 \times 0.667$ Pa $= 6.148$ kPa
$F = pA_i = 6.148 \times A(0.3)$ kN $= \textbf{435 N}$
Using Equation 10.2 with inclined dimensions $d_i = 0.3$ m (same as d) and $h_i = 1.0$ m

$$y_i = \frac{d_i^2}{16h_i} = \frac{0.3^2}{16 \times 1.0} = 0.0056 \text{ m} = 5.6 \text{ mm}$$

Therefore the force acts at a distance of $200 - 5.6 = \textbf{194.4 mm}$ upward along the slope from the base.

10.6 (a) Weight of water displaced $= \rho V g = 0.024 \times A(0.2) \times 1030 \times 9.81 = 7.619$ N
$\therefore$ Weight of block $= 7.619$ N $\therefore$ mass of block $= \textbf{0.777 kg}$

Volume of block $= 0.1^3 = 1 \times 10^{-3}$ m^3 $\therefore$ density $= \dfrac{m}{V} = \dfrac{0.777}{1 \times 10^{-3}} = \textbf{777 kg/m}^3$

(b) Fully submerged, buoyancy force = weight of fluid displaced
$$= \rho Vg = 1030 \times 1 \times 10^{-3} \times 9.81 = 10.104 \text{ N}$$
∴ Force needed to submerge block $= 10.104 - 7.619 = \textbf{2.49 N}$

Chapter 11

11.1 **(a)** $\dot{V} = \dfrac{V}{t} = \dfrac{25 \times 10^{-3}}{4 \times 60 + 24}$ m³/s = **0.0947 L/s**

$$v = \frac{\dot{V}}{A} = \frac{0.0947 \times 10^{-3}}{A(0.01)} = \textbf{1.21 m/s}$$

(b) $\dot{V} = vA$
$$= 5.6 \times A(0.2)$$
$$= \textbf{0.176 m}^3\textbf{/s}$$
$\dot{m} = \rho\dot{V}$
$$= 0.9 \times 10^3 \times 0.176$$
$$= \textbf{158 kg/s}$$

11.2 Referring to Figure S7, let $\dot{m}_\text{L}$ be the mass-flow rate of lost steam.

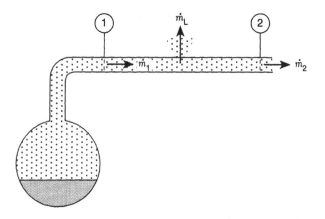

Fig. S7

$$\dot{m}_1 = 12.6 \text{ t/h} = \frac{12\,600}{3600} = 3.5 \text{ kg/s}$$

and $\dot{m}_2 = v_2 A_2 \rho_2$

but $\rho = \dfrac{1}{v}$ (v = specific volume)

∴ $\dot{m}_2 = 25 \times A(0.25) \times \dfrac{1}{0.365}$
$$= 3.362 \text{ kg/s}$$

From the equation of continuity:
$$\dot{m}_1 = \dot{m}_\text{L} + \dot{m}_2$$
∴ $\dot{m}_\text{L} = 3.5 - 3.362 = 0.138 \text{ kg/s}$
$$= 496 \text{ kg/h}$$

∴ percentage loss $= \dfrac{496}{12\,600} \times 100 = \textbf{3.94\%}$

NEW 11/8/97

11.3 Pressure head $h_p = \dfrac{p}{\rho g} = \dfrac{30 \times 10^3}{0.9 \times 10^3 \times 9.81} =$ **3.4 m**

Velocity head $h_v = \dfrac{v^2}{2g} = \dfrac{5^2}{2 \times 9.81} =$ **1.27 m**

Potential head $h =$ **3.5 m**

Total head $= 3.4 + 1.27 + 3.5 =$ **8.17 m**

11.4 (a) $\dot{m}_1 = v_1 A_1 \rho_1 \therefore v_1 = \dfrac{\dot{m}_1}{A_1 \rho_1} = \dfrac{40}{A(0.1) \times 900} =$ **5.66 m/s**

(b) $h_{v_1} = \dfrac{v_1^2}{2g} = \dfrac{5.66^2}{19.62} =$ **1.632 m**

(c) $h_{p_1} = \dfrac{p_1}{\rho g} = \dfrac{150 \times 10^3}{900 \times 9.81} =$ **17 m**

(d) $h_1 = h_2$ also $h_{p_2} = 0$ (atmospheric pressure).
Bernoulli equation simplifies to:
$h_{p_1} + h_{v_1} = h_{v_2}$
$17 + 1.632 = h_{v_2} \therefore h_{v_2} =$ **18.62 m**

(e) $h_{v_2} = \dfrac{v_2^2}{2g}$
$\therefore v_2^2 = 18.62 \times 19.62 \quad \therefore v_2 =$ **19.11 m/s**

(f) By continuity: $v_1 d_1^2 = v_2 d_2^2$
$\therefore d_2^2 = \dfrac{v_1 d_1^2}{v_2} = \dfrac{5.66 \times 0.1^2}{19.11} \quad \therefore d_2 = 0.0544$ m $=$ **54.4 mm**

11.5 (a) Applying Bernoulli's equation between ① and ② with:
$p_1 = 0$ (atmospheric), $v_1 = 0$ (large tank), $h_1 = 3.5$ m, $p_2 = 0$ (atmospheric), $h_2 = 2.2$ m
$\therefore h_1 = \dfrac{v_2^2}{2g} + h_2 \quad \therefore v_2 = \sqrt{2g(h_1 - h_2)}$

Note This is Torricelli's equation using $h = h_1 - h_2$
$\therefore v_2 = \sqrt{19.62 \times (3.5 - 2.2)} =$ **5.05 m/s**

(b) $\dot{V} = vA = 5.05 \times A(0.01)$ m³/s $=$ **0.397 L/s**

(c) Applying Bernoulli's equation between ① and ③ with:
$p_1 = 0$ (atmospheric), $v_1 = 0$ (large tank), $h_1 = 3.5$ m, $h_3 = 4.8$ m, $v_3 = 5.05 \times \left(\dfrac{10}{12}\right)^2$
$= 3.51$ m/s

$$h_1 = \dfrac{p_3}{\rho g} + \dfrac{v_3^2}{2g} + h_3$$

$\therefore 3.5 = \dfrac{p_3}{10^3 \times 9.81} + \dfrac{3.51^2}{2 \times 9.81} + 4.8$

$\therefore p_3 =$ **−18.9 kPa** (the negative sign indicating pressure below atmospheric or vacuum)

Note The same result could be obtained by applying Bernoulli's equation between ② and ③.

11.6 (a) $h_1 = 2.5$ m, $h_2 = 0.5$ m

(b) $\dot{m} = vA\rho \therefore v_1 = \dfrac{\dot{m}}{A_1\rho} = \dfrac{9.5}{A(0.1) \times 1.15 \times 10^3} = 1.052$ m/s

$$h_{v1} = \frac{v_1^2}{2g} = \frac{1.052^2}{19.62} = 0.0564 \text{ m}$$

$$v_2 = v_1 \times \left(\frac{\phi_1}{\phi_2}\right)^2 = 1.052 \times 4 = 4.207 \text{ m/s}$$

$$h_{v2} = \frac{v_2^2}{2g} = \frac{4.207^2}{19.62} = 0.9022 \text{ m}$$

(c) $h_{p2} = \dfrac{p_2}{\rho g} = \dfrac{30 \times 10^3}{1.15 \times 10^3 \times 9.81} = 2.659$ m

Applying Bernoulli's equation between ① and ②:

$h_{p1} + h_{v_1} + h_1 = h_{p2} + h_{v_2} + h_2$

$\therefore h_{p1} + 0.0564 + 2.5 = 2.659 + 0.9022 + 0.5$

$\therefore h_{p1} + 2.5564 = 4.061$ (total head)

$\therefore h_{p1} = \mathbf{1.505}$ **m**

(d) These heads may now be graphed as shown in Figure S8:

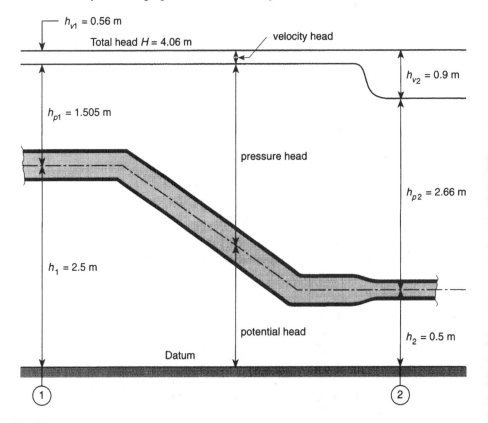

Fig. S8

11.7 (a) $v_1 = 50 \times \left(\dfrac{5}{12}\right)^2 = 8.68$ m/s

$$\frac{p_1}{\rho g} + \frac{v_1^2}{2g} + \cancel{h_1} = \frac{\cancel{p_2}}{\cancel{\rho g}} + \frac{v_2^2}{2g} + h_2$$

$$\therefore \frac{p_1}{\rho g} + \frac{8.68^2}{19.62} = \frac{50^2}{19.62} + 1$$

$$\therefore \frac{p_1}{\rho g} = 124.6 \text{ m}$$

$$\therefore p_1 = \mathbf{1.22\ MPa}$$

(b) If losses are 10 m,

$$\frac{p_1}{\rho g} = 124.6 + 10$$

$$= 134.6 \text{ m}$$

$$\therefore p_1 = \mathbf{1.32\ MPa}$$

Chapter 12

12.1 $\dot{m} = vA\rho = 2.2 \times A(0.065) \times 1.1 \times 10^3 = 8.03$ kg/s

$H = 26.3 - 3.6 = 22.7$ m

$P_f = \dot{m}gH = 8.03 \times 9.81 \times 22.7$ W $= \mathbf{1.79\ kW}$

12.2 (a) $\dot{V} = vA = 8.3 \times A(0.5) = \mathbf{1.63\ m^3/s}$

(b) $\rho_{air} = 1/0.83 = 1.205$ kg/m^3

$\dot{m} = vA\rho = \dot{V}\rho = 1.63 \times 1.205 = \mathbf{1.964\ kg/s}$

(c) $p = \rho_{water}\, gh = 10^3 \times 9.81 \times (0.126 - 0.02)$ Pa $= \mathbf{1.04\ kPa}$

(d) $H = h_p = \dfrac{p}{\rho_{air}g} = \dfrac{1.04 \times 10^3}{1.205 \times 9.81} = \mathbf{88\ m}$

(e) $P_f = \dot{m}gH = 1.964 \times 9.81 \times 88$ (W) $= \mathbf{1.69\ kW}$

(f) $P_f = p\dot{V} = 1.04 \times 10^3 \times 1.63$ (W) $= \mathbf{1.69\ kW}$

12.3 (a) The difference in mercury height is 9 mm = 0.009 m

The difference in pressure in the pipe is given by:

$p = \rho_{mercury}\, gh = 13.6 \times 10^3 \times 9.81 \times 0.009$ Pa $= \mathbf{1.2\ kPa}$

(b) $H_L = \dfrac{p}{\rho g} = \dfrac{1.2 \times 10^3}{1.05 \times 10^3 \times 9.81} = \mathbf{0.1166\ m}$

(c) $\dot{V} = vA = 2.8 \times A(0.25) = 0.1374$ m^3/s

$\dot{m} = vA\rho = \dot{V}\rho = 0.1374 \times 1.05 \times 10^3 = 144.3$ kg/s

Using Equation 12.1 $P_f = \dot{m}gH = 144.3 \times 9.81 \times 0.1166 = \mathbf{165\ W}$

Using Equation 12.2 $P_f = p\dot{V} = 1.2 \times 10^3 \times 0.1374 = \mathbf{165\ W}$

12.4 $\dot{V} = vA = 3.2 \times A(0.1) = 0.0251$ m^3/s

$$\eta = \frac{P_f}{P} \therefore P_f = P \times \eta = 15 \times 0.65 = 9.75 \text{ kW}$$

Using Equation 12.2: $P_f = p\dot{V} \therefore p = \dfrac{P_f}{\dot{V}} = \dfrac{9.75 \times 10^3}{0.0251}$ Pa $= 388$ kPa

This is the pressure change *across* the pump

$\therefore$ pressure at the outlet flange $= 388 - 22 = \mathbf{366\ kPa}$

12.5 Denoting the water surface ① and the exit position ②:

$p_1 = 0$ (surface at atmospheric pressure), $v_1 = 0$ (large reservoir)

$p_2 = 0$ (discharge to atmospheric pressure), $h_2 = 0$ (datum)

Substituting in Equation 12.5 with H negative (energy out of the system):

$$\frac{\cancel{p_1}}{\cancel{\rho g}} + \frac{\cancel{v_1^2}}{\cancel{2g}} + h_1 - H = \frac{\cancel{p_2}}{\cancel{\rho g}} + \frac{v_2^2}{2g} + \cancel{h_2} + H_L$$

$$\therefore 9.3 - H = \frac{3.5^2}{19.62} + 0.6$$

$\therefore H = 8.0756$ m

Now $P_f = \dot{m}gH = 10^3 \times 9.81 \times 8.0756 = 79.22$ kW

But $P = \eta P_f = 0.92 \times 79.22 = \textbf{72.9 kW}$

Chapter 13

13.1 **(a)** $\dot{m} = vA\rho = 560 \times A(0.35) \times 0.55 = 29.63$ kg/s

$v_2 = 560$ m/s, $v_1 = 0$ (there is no intake velocity in a rocket motor)

$F = \dot{m}(v_2 - v_1) = 29.63 \times (560 - 0) = \textbf{16.6 kN}$

(b) Weight of rocket $= 1.2 \times 9.81 = 11.77$ kN

$\therefore$ Nett upward force $= 16.6 - 11.77 = 4.82$ kN (4823 N)

$F = ma \therefore 4823 = 1200a$

$\therefore a = \textbf{4.02 m/s}^2$

13.2 $\dot{m} = vA\rho = 50 \times A(0.005) \times 1.2 = 1.178 \times 10^{-3}$ kg/s

$F = \dot{m}(v_2 - v_1) = 1.178 \times 10^{-3} \times (0 - 50) = -58.9 \times 10^{-3}$ N (on fluid ←)

$\therefore$ force on the surface is $\textbf{58.9} \times \textbf{10}^{-3}$ **N** →

13.3 Refer to Figure S9.

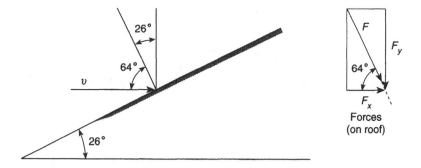

Fig. S9

Now $v = 45$ km/h $= \dfrac{45}{3.6}$ m/s $= 12.5$ m/s

And $\dot{m} = vA\rho = 12.5 \times A(0.035) \times 1000 = 12.026$ kg/s

In the perpendicular direction, $v_1 = v \sin 26° = 5.48$ m/s and $v_2 = 0$

The force on the jet is $F = \dot{m}(v_2 - v_1) = 12.026 \times (0 - 5.48) = -65.9$ N

Because there is no force parallel to the roof, the resultant force on the roof is $\textbf{65.9 N}$ acting downward at an angle of $\textbf{64°}$ (that is an angle of $\textbf{-64°}$).

In the horizontal (x direction), the force on the roof is:
$Fx = 65.9 \cos -64° = $ **28.9 N** $\rightarrow$
In the vertical (y direction), the force on the roof is:
$Fy = 65.9 \sin -64°, = $ **59.2 N** $\downarrow$

13.4 $\dot{m} = vA\rho = 20 \times A(0.012) \times 1000 = 2.2619$ kg/s
(a) Using the y direction: $v_1 = 20$ m/s $\uparrow$ $v_2 = -20 \cos 20° = -18.79$ m/s $\downarrow$
  The force on the water is: $F = \dot{m}(v_2 - v_1) = 2.2619 \times (-18.79 - 20) = -87.75$ N
  The negative sign indicates that the force on the water is acting downward $\downarrow$
  $\therefore$ the force on the cup is **87.75 N** $\uparrow$ (there is no force in the x direction)
(b) If friction causes a reduction in velocity of 5%, then in the y direction:
  $v_2 = -18.79 \times 0.95 = -17.85$ m/s $\downarrow$ and v_1 is the same (20 m/s) $\uparrow$
  The force on the water is: $F = \dot{m}(v_2 - v_1) = 2.2619 \times (-17.85 - 20) = -85.6$ N $\downarrow$
  $\therefore$ the force on the cup is **85.6 N** $\uparrow$

13.5 The relative velocity of the water to the cup is now 15 m/s (20 – 5)
  $\dot{m} = vA\rho = 15 \times A(0.012) \times 1000 = 1.6965$ kg/s
(a) Using the y direction, the initial relative velocity of the water to the cup is:
  $v_1 = 15$ m/s $\uparrow$ and the final relative velocity is: $v_2 = -15 \cos 20° = -14.095$ m/s $\downarrow$
  The force on the water is: $F = \dot{m}(v_2 - v_1) = 1.6965 \times (-14.095 - 15) = -49.36$ N $\downarrow$
  $\therefore$ the force on the cup is **49.4 N** $\uparrow$
(b) If friction causes a reduction in velocity of 5%, then in the y direction:
  $v_2 = -14.095 \times 0.95 = -13.39$ m/s $\downarrow$ and v_1 is the same (15 m/s) $\uparrow$
  The force on the water is: $F = \dot{m}(v_2 - v_1) = 1.6965 \times (-13.39 - 15) = -48.16$ N $\downarrow$
  $\therefore$ the force on the cup is **48.2 N** $\uparrow$

13.6 (a) $v_2 = 5$ m/s $v_1 = 5 \times \left(\dfrac{45}{30}\right)^2 = 11.25$ m/s

Apply Bernoulli's equation between ① and ②:

$$\frac{p_1}{\rho g} + \frac{v_1^2}{2g} + h_1 = \frac{p_2}{\rho g} + \frac{v_2^2}{2g} + h_2$$

But $h_1 = h_2$ (horizontal) and g is now common to all terms so divide by g:

$$\frac{p_1}{\rho} + \frac{v_1^2}{2} = \frac{p_2}{\rho} + \frac{v_2^2}{2}$$

$$\therefore \frac{p_1}{1000} + \frac{11.25^2}{2} = \frac{200 \times 10^3}{1000} + \frac{5^2}{2}$$

$\therefore p_1 = $ **149.2 kPa**

(b) Calculating mass-flow rate from downstream conditions:
  $\dot{m} = vA\rho = 5 \times A(0.045) \times 1000 = 7.952$ kg/s
  Assuming that the nett force on the fluid is $+$ ($\rightarrow$)
  $p_1A_1 - p_2A_2 + F = \dot{m}(v_2 - v_1)$
  $\therefore 149.2 \times 10^3 \times A(0.03) - 200 \times 10^3 \times A(0.045) + F = 7.952 \times (5 - 11.25)$
  $\therefore 105.5 - 318.1 + F = -49.7$
  $\therefore F = 163$ N ($\rightarrow$ on fluid)
  $\therefore F = $ **-163 N** ($\leftarrow$ on transition)

Appendixes

Appendix 1: Principal symbols

Notes A dot above a symbol means 'per second', that is a flow rate. 'Specific' means 'per kg' (the only exception being specific fuel consumption).

Where the same symbol is used for several different quantities, the meaning of the symbol should be clear from the context. If in doubt when using an equation *check units*.

Symbol	Quantity	Base unit (or value)
A	area	m²
a	acceleration	m/s²
c	specific heat capacity	J/kgK
c	speed of light	300 000 km/s
c_p	specific heat capacity at constant pressure	J/kgK
c_v	specific heat capacity at constant volume	J/kgK
E	energy content of a fuel	J/kg
E	nuclear energy	J
F	force	N
FP	friction power	W
g	gravitational constant	9.81 m/s²; 9.81 N/kg
H	enthalpy	J
H	height of a rectangular opening (in a fluid)	m
H	total head (of a fluid)	m
H_L	head loss (of a fluid)	m
h	specific enthalpy	J/kg
h	vertical height	m
h	potential head (of a fluid)	m
h_p	pressure head (of a fluid)	m
h_v	velocity head (of a fluid)	m
IP	indicated power	W
k_s	spring constant	Pa/m
KE	kinetic energy	J
L	latent heat	J/kg
L	piston stroke	m
l	length	m
$\ln$	natural logarithm (to base e)	—
m	mass	kg
M	relative molecular mass	—
N	rotational speed	rpm
n	cycles per second	s⁻¹
n	polytropic index (index of compression or expansion)	—
P	power (shaft or output)	W
P_f	fluid power	W

Symbol	Quantity	Base unit (or value)
p	pressure	Pa
p_m	mean effective pressure	Pa
PE	potential energy	J
Q	heat energy	J
Q_S	heat supplied	J
Q_R	heat rejected	J
R	gas constant	J/kgK
RD	relative density	—
r	radius	m
T	temperature	K
T	torque	Nm
t	time	s
U	internal energy	J
u	specific internal energy	J/kg
V	volume	m³
v	specific volume	m³/kg
W	work	J
w_t	weight	N
x	displacement	m
y	centre of pressure distance (from a liquid surface)	m

Greek symbols

Symbol	Quantity	Base unit (or value)
θ	angular displacement	rad
ω	angular velocity	rad/s
γ	adiabatic index (c_p/c_v)	—
Δp	change in pressure	Pa
ΔT	change in temperature	K
ρ	density	kg/m³
μ	dynamic viscosity	Pas
η	efficiency (overall)	—
η_m	mechanical efficiency	—
η_{ind}	indicated thermal efficiency	—
ν	kinematic viscosity	m²/s
υ	velocity	m/s

Appendix 2: Principal equations

Notes Before using an equation it is good practice to check units for dimensional consistency. Subscript 1 is used for initial conditions and subscript 2 for final conditions.

2.1 $\rho = \dfrac{m}{V}$ density

2.2 $RD \text{ (substance)} = \dfrac{\rho \text{ (substance)}}{\rho \text{ (water)}}$ relative density

2.3 $v = \dfrac{V}{m} = \dfrac{1}{\rho}$ specific volume

2.4 $F = ma$ force

2.5 $w_t = mg$ weight

2.6 $p = \dfrac{F}{A}$ pressure

2.7 $p = p_g + p_{atm}$ absolute pressure

2.8 $p = \rho gh$ pressure increase due to self-weight of a liquid

2.9 $T(K) = 273 + T(°C)$ absolute temperature

2.10 Compression ratio $= \dfrac{V_1}{V_2} = \dfrac{V_s + V_c}{V_c}$ compression ratio

2.11 Pressure ratio $= \dfrac{p_2}{p_1}$ pressure ratio

3.1 $\eta = \dfrac{\text{usable energy output}}{\text{energy input}}$ efficiency of energy conversion

$\dfrac{\text{energy input} - \text{energy loss}}{\text{energy input}}$

3.2 $PE = mgh$ potential energy

3.3 $KE = \frac{1}{2}mv^2$ linear kinetic energy

3.4 $W = Fx$ linear work

3.5 $W = T\theta$ rotational work

3.6 $W = pV = p(V_2 - V_1)$ work for a constant-pressure process

3.7 $W = \displaystyle\int_{V_1}^{V_2} p\,dV$ work for a variable-pressure process

3.8 $P = \dfrac{W}{t} = Fv$ power (linear)

3.9 $P = T\omega$ power (rotational)

3.10 $\omega = \dfrac{2\pi N}{60} = \dfrac{\pi N}{30}$ angular velocity

3.11 $Q = mc\Delta T = mc(T_2 - T_1)$ sensible heat

3.12 $Q = mL$ latent heat

3.13 $Q = mE$ combustion heat energy

3.14 $E = mc^2$ nuclear energy

4.1 $Q - W = U_2 - U_1$ non-flow energy equation

4.2 $U_2 - U_1 = mc(T_2 - T_1)$ change in internal energy

4.3 $\sum \dot{m}_{in} = \sum \dot{m}_{out} = \text{constant}$ continuity equation for mass flow

4.4 $\dot{V} = \dfrac{V}{t} = vA$ volume-flow rate

4.5 $\dot{m} = \dfrac{m}{t} = vA\rho$ mass-flow rate

4.6 $\sum \dot{V}_{in} = \sum \dot{V}_{out} = \text{constant}$ continuity equation with no density change

4.7 $H = U + pV$ enthalpy

4.8 $Q - W = H_2 - H_1$ steady-flow energy equation

4.9 $H_2 - H_1 = m(h_2 - h_1)$ enthalpy change

4.10 $h_2 - h_1 = c_p(T_2 - T_1)$ specific enthalpy change for a gas

5.1 $\dfrac{pV}{T} = c$ or $\dfrac{p_1 V_1}{T_1} = \dfrac{p_2 V_2}{T_2}$ general gas equation

5.2 $pV = mRT$ equation of state

5.3 $R = \dfrac{8314}{M}$ gas constant

5.4 $Q = mc_p(T_2 - T_1)$ constant-pressure heating

5.5 $Q = mc_v(T_2 - T_1)$ constant-volume heating

5.6 $R = c_p - c_v$ gas constant

5.7 $U_2 - U_1 = mc_v(T_2 - T_1)$ internal energy change

5.8 $H_2 - H_1 = mc_p(T_2 - T_1)$ enthalpy change

5.9 $\dfrac{V_1}{T_1} = \dfrac{V_2}{T_2}$ constant-pressure process

5.10 $\dfrac{p_1}{T_1} = \dfrac{p_2}{T_2}$ constant-volume process

5.11 $p_1 V_1 = p_2 V_2$ isothermal process

5.12 $W = p_1 V_1 \ln\left(\dfrac{V_2}{V_1}\right)$ isothermal work

5.13 $\quad p_1 V_1^n = p_2 V_2^n$ $\hfill$ polytropic process

5.14 $\quad \dfrac{T_2}{T_1} = \left(\dfrac{V_1}{V_2}\right)^{n-1}$ $\hfill$ polytropic process

5.15 $\quad \dfrac{p_2}{p_1} = \left(\dfrac{T_2}{T_1}\right)^{\frac{n}{n-1}}$ $\hfill$ polytropic process

5.16 $\quad W = \dfrac{p_1 V_1 - p_2 V_2}{n - 1}$ $\hfill$ polytropic work

5.17 $\quad \gamma = \dfrac{c_p}{c_v}$ $\hfill$ adiabatic index

6.1 $\quad W = Q_S - Q_R$ $\hfill$ work output of a heat engine

6.2 $\quad P = \dot{Q}_S - \dot{Q}_R$ $\hfill$ power output of a heat engine

6.3 $\quad \eta = \dfrac{W}{Q_S} = \dfrac{P}{\dot{Q}_S}$ $\hfill$ efficiency of a heat engine

6.4 $\quad \eta = 1 - \dfrac{Q_R}{Q_S} = 1 - \dfrac{\dot{Q}_R}{\dot{Q}_S}$ $\hfill$ efficiency of a heat engine

6.5 $\quad \eta_C = 1 - \dfrac{T_c}{T_h}$ $\hfill$ Carnot efficiency

6.6 $\quad \eta = 1 - \dfrac{1}{\left(\dfrac{V_1}{V_2}\right)^{\gamma-1}}$ $\hfill$ theoretical efficiency of Otto cycle

7.1 $\quad \dot{Q}_S = \dot{m}E$ $\hfill$ heat-supply rate

7.2 $\quad p_m = \dfrac{A_c k_s}{l_c}$ $\hfill$ mean effective pressure

7.3 $\quad IP = p_m L A n$ $\hfill$ indicated power

7.4 $\quad FP = IP - P$ $\hfill$ friction power

7.5 $\quad \eta_m = \dfrac{P}{IP}$ $\hfill$ mechanical efficiency

7.6 $\quad \eta_{ind} = \dfrac{IP}{\dot{Q}_S}$ $\hfill$ indicated thermal efficiency

8.1 $\quad v = \dfrac{\mu}{\rho}$ $\hfill$ kinematic viscosity

10.1 $y = \dfrac{H^2}{12h}$ centre of pressure distance for a rectangle

10.2 $y = \dfrac{d^2}{16h}$ centre of pressure distance for a circle

11.1 $h_p = \dfrac{p}{\rho g}$ pressure head

11.2 $h_v = \dfrac{v^2}{2g}$ velocity head

11.3 $H = h_p + h_v + h$ total head

11.4 $\dfrac{p_1}{\rho g} + \dfrac{v_1^2}{2g} + h_1 = \dfrac{p_2}{\rho g} + \dfrac{v_2^2}{2g} + h_2$ Bernoulli equation

11.5 $\Delta p = \rho g H_L$ pressure drop due to head loss

11.6 $\dfrac{p_1}{\rho g} + \dfrac{v_1^2}{2g} + h_1 = \dfrac{p_2}{\rho g} + \dfrac{v_2^2}{2g} + h_2 + H_L$ Bernoulli equation with head loss

12.1 $P_f = \dot{m}gH$ fluid power

12.2 $P_f = p\dot{V}$ fluid power due to a pressure change

12.3 $\eta = \dfrac{P_f}{P}$ efficiency of a pump or compressor

12.4 $\eta = \dfrac{P}{P_f}$ efficiency of a turbine or motor

12.5 $\dfrac{p_1}{\rho g} + \dfrac{v_1^2}{2g} + h_1 \pm H = \dfrac{p_2}{\rho g} + \dfrac{v_2^2}{2g} + h_2 + H_L$ Bernoulli equation with fluid machinery

13.1 $F = \dot{m}(v_2 - v_1)$ fluid dynamics force

Appendix 3: Approximate relative atomic and molecular masses of some elements

Element	Chemical symbol	Relative atomic mass	Molecular symbol	Relative molecular mass
hydrogen	H	1	H_2	2
helium	He	4	He	4
carbon	C	12	C	12
nitrogen	N	14	N_2	28
oxygen	O	16	O_2	32
sulphur	S	32	S	32
argon	Ar	40	Ar	40

Appendix 4: Specific heat capacity of some substances (medium-temperature range)

Substance	Specific heat capacity (kJ/kgK)
air (constant pressure)	1.005
(constant volume)	0.718
aluminium (and alloys)	0.88
copper and brass	0.39
cast iron	0.42
glass	0.84
steel (and alloys)	0.46
lead	0.13
mercury	0.14
water	4.19
timber (Oregon)	2.6
ice	2.04

Appendix 5: Pressure–height relationship

Consider a substance of constant density ρ, with area A and height h as shown in Figure A5.

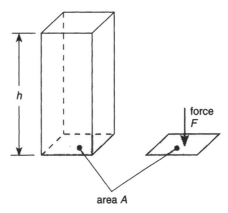

force
F

h

area A

Fig. A5

The volume of substance enclosed is: Ah

The mass of substance enclosed is: ρAh

The force acting on area A = weight of substance enclosed = $mg = \rho Ahg$

The pressure on area $A = \dfrac{F}{A} = \dfrac{\rho Ahg}{A} = \rho gh$

Therefore $p = \rho gh$ (which is Equation 2.8)

Note   For simplicity, area A is shown in Figure A5 as being square, but this area can be of any shape since this does not affect the validity of the proof.

Appendix 6: Characteristic gas constant

Consider a constant-pressure process applied to a mass of gas (ideal).
The non-flow energy equation may be used since this applies to *any* process on a fixed mass of gas:

$$Q - W = (U_2 - U_1)$$

Now for a constant-pressure process:

$$W = p(V_2 - V_1)$$
$$Q = mc_p(T_2 - T_1)$$
$$U_2 - U_1 = mc_v(T_2 - T_1)$$

Substituting,

$$mc_p(T_2 - T_1) - p(V_2 - V_1) = mc_v(T_2 - T_1)$$

Since $p = p_1 = p_2$
$$p(V_2 - V_1) = p_2V_2 - p_1V_1$$

But $p_2V_2 = mRT_2$ and $p_1V_1 = mRT_1$
$$\therefore p(V_2 - V_1) = mR(T_2 - T_1)$$

Substituting:
$$mc_p(T_2 - T_1) - mR(T_2 - T_1) = mc_v(T_2 - T_1)$$
$$\therefore c_p - R = c_v$$

or

$$\boxed{R = c_p - c_v}$$

Which is Equation 5.6

Note   Although Equation 5.6 was derived by considering a constant-pressure process, R, c_p and c_v are all properties of a gas and do not change their values despite the process for any given ideal gas. Hence Equation 5.6 is of general validity and describes the fundamental relationship between these properties. For example, R for air is 287 J/kgK. Now c_p for air is 1005 J/kgK and c_v is 718 J/kgK. It is indeed evident that $R = c_p - c_v$.

Appendix 7: Isothermal work

Now $\quad W = \int_{V_1}^{V_2} p\,dV$

For an isothermal process:

$$pV = c \therefore p = \frac{c}{V}$$

$$\therefore W = \int_{V_1}^{V_2} \frac{c}{V}\,dV$$

$$= c[\ln V]_{V_1}^{V_2}$$

$$= c(\ln V_2 - \ln V_1)$$

$$= c \ln \left(\frac{V_2}{V_1}\right)$$

But $\quad c = pV = p_1V_1 = p_2V_2$

$$\boxed{W = p_1V_1\ln \left(\frac{V_2}{V_1}\right)}$$

which is Equation 5.12

Appendix 8: Polytropic pressure–volume–temperature relationships

For a polytropic process:

$$pV^n = c$$

$$\therefore \qquad \boxed{p_1 V_1{}^n = p_2 V_2{}^n} \qquad \text{polytropic process (5.13)}$$

For *any* process on a fixed mass of ideal gas:

$$\frac{p_1 V_1}{T_1} = \frac{p_2 V_2}{T_2}$$

$$\therefore \frac{p_2}{p_1} = \frac{V_1}{V_2} \times \frac{T_2}{T_1} \quad \dots (1)$$

From Equation 5.13:

$$\frac{p_2}{p_1} = \frac{V_1{}^n}{V_2{}^n} = \left(\frac{V_1}{V_2}\right)^n$$

Substituting in (1):

$$\frac{V_1}{V_2} \times \frac{T_2}{T_1} = \left(\frac{V_1}{V_2}\right)^n$$

$$\therefore \frac{T_2}{T_1} = \left(\frac{V_1}{V_2}\right)^n \times \frac{V_2}{V_1}$$

$$= \left(\frac{V_1}{V_2}\right)^n \times \left(\frac{V_1}{V_2}\right)^{-1}$$

$$\therefore \qquad \boxed{\frac{T_2}{T_1} = \left(\frac{V_1}{V_2}\right)^{n-1}} \qquad \text{polytropic process (5.14)}$$

This equation may also be written as:

$$\frac{V_1}{V_2} = \left(\frac{T_2}{T_1}\right)^{\frac{1}{n-1}}$$

Substituting in (1):

$$\frac{p_2}{p_1} = \left(\frac{T_2}{T_1}\right)^{\frac{1}{n-1}} \times \frac{T_2}{T_1}$$

$$= \left(\frac{T_2}{T_1}\right)^{1+\frac{1}{n-1}}$$

$\therefore$

$$\boxed{\frac{p_2}{p_1} = \left(\frac{T_2}{T_1}\right)^{\frac{n}{n-1}}}$$ polytropic process (5.15)

Note   Equations 5.13, 5.14 and 5.15 may be used for any possible combination of relationships between p and V, T and V or p and T.

Appendix 9: Polytropic work

Now $\qquad W = \int_{V_1}^{V_2} p dV$

Since $\quad pV^n = c$, for a polytropic process

$$\therefore p = \frac{c}{V^n}$$

Substituting:

$$W = \int_{V_1}^{V_2} \frac{c}{V^n} dV$$

$$= \int_{V_1}^{V_2} cV^{-n} dV$$

$$= \frac{1}{-n+1} [cV^{-n+1}]_{V_1}^{V_2}$$

$$= \frac{1}{-n+1} c(V_2^{-n+1} - V_1^{-n+1})$$

Now $\qquad c = p_1V_1^n = p_2V_2^n$. Therefore

$$W = \frac{1}{-n+1} (p_2V_2^n V_2^{-n+1} - p_1V_1^n V_1^{-n+1})$$

$$= \frac{1}{-n+1} (p_2V_2 - p_1V_1)$$

Multiplying top and bottom by -1:

$$\boxed{W = \frac{p_1V_1 - p_2V_2}{n-1}} \qquad \text{work transfer for a polytropic process (5.16)}$$

Notes

1 Since $P_1V_1 - P_2V_2 = mR(T_1 - T_2)$, an alternative form is:

$$W = \frac{mR}{n-1} (T_1 - T_2)$$

2 For an *expansion* process, $T_1 > T_2$, and for a *compression* process, $T_2 > T_1$, so the work will be positive for an expansion process and negative for a compression process.

Appendix 10: Adiabatic index

Since $Q - W = U_2 - U_1$, then $Q = U_2 - U_1 + W$

For a polytropic process, $W = \dfrac{p_1V_1 - p_2V_2}{n - 1}$ and $U_2 - U_1 = mc_v(T_2 - T_1)$

Substituting,

$$Q = mc_v(T_2 - T_1) + \frac{p_1V_1 - p_2V_2}{n - 1}$$

$$= -mc_v(T_1 - T_2) + \frac{p_1V_1 - p_2V_2}{n - 1}$$

This expression may be simplified. Since $p_1V_1 - p_2V_2 = mR(T_1 - T_2)$, then

$$Q = -mc_v(T_1 - T_2) + \frac{mR}{n - 1}(T_1 - T_2)$$

$$= m(T_1 - T_2)\left(\frac{R}{n - 1} - c_v\right)$$

Also, since $\qquad R = c_p - c_v$ this expression reduces to:

$$Q = m(T_1 - T_2)\frac{c_p - nc_v}{n - 1}$$

For the adiabatic process, $Q = 0$. Hence

$$\frac{m(c_p - nc_v)}{n - 1}(T_1 - T_2) = 0$$

For a polytropic process $(T_1 - T_2)$ cannot be equal to 0 (as this would make the process isothermal), this means that

$$\frac{m(c_p - nc_v)}{n - 1} = 0$$

or

$$c_p = nc_v$$

$$\therefore n = \frac{c_p}{c_v}$$

The ratio $\dfrac{c_p}{c_v}$ occurs frequently; it is called the **adiabatic index** and is given the symbol γ.

$$\boxed{\gamma = \frac{c_p}{c_v}}$$
adiabatic index (5.17)

Note Since c_p and c_v are constant for ideal gases, γ must also be constant, Hence the adiabatic process may be treated as a special case of the polytropic process in which the index $n = \gamma = c_p/c_v$. For example, for air (since $c_p = 1005$ J/kgK and $c_v = 718$ J/kgK), $\gamma = 1.4$.

Appendix 11: Carnot efficiency

The following processes constitute the Carnot cycle (refer to Figure 6.9):

① → ②   isothermal compression; heat rejected to sink at constant temperature T_c; the pressure increases

② → ③   adiabatic compression; the pressure and temperature both increase without any heat flow taking place

③ → ④   isothermal expansion; heat supplied from source at constant temperature T_h; the pressure decreases

④ → ①   adiabatic expansion; the pressure and temperature both decrease until they have their same initial value; no heat flow occurs

For an isothermal compression of an ideal gas:

$$Q_R = W_{① \to ②} = p_1 V_1 \ln \left(\frac{V_1}{V_2} \right)$$

Note This formula has been written $p_1 V_1 \ln \left(\frac{V_1}{V_2} \right)$ rather than $p_1 V_1 \ln \left(\frac{V_2}{V_1} \right)$ to avoid the negative term. (See note 2 in Section 6.3.)

For an isothermal expansion of an ideal gas:

$$Q_S = W_{③ \to ④} = p_3 V_3 \ln \left(\frac{V_4}{V_3} \right)$$

Also, $p_1 V_1 = mRT_1$ and $p_3 V_3 = mRT_3$

$$\therefore Q_S = mRT_3 \ln \left(\frac{V_4}{V_3} \right) = mRT_h \ln \left(\frac{V_4}{V_3} \right)$$

and, $$Q_R = mRT_1 \ln \left(\frac{V_1}{V_2} \right) = mRT_c \ln \left(\frac{V_1}{V_2} \right)$$

From Equation 6.4:

$$\eta = 1 - \frac{Q_R}{Q_S} = \eta_C \text{ (Carnot efficiency)}$$

$$\therefore \eta_C = 1 - \frac{T_c \ln \left(\frac{V_1}{V_2} \right)}{T_h \ln \left(\frac{V_4}{V_3} \right)}$$

But for ② → ③ (adiabatic compression):

$$\frac{T_3}{T_2} = \left(\frac{V_2}{V_3} \right)^{\gamma - 1} = \frac{T_h}{T_c}$$

322

and for ④ → ① (adiabatic expansion):

$$\frac{T_4}{T_1} = \left(\frac{V_1}{V_4}\right)^{\gamma-1} = \frac{T_h}{T_c}$$

$$\therefore \quad \frac{V_1}{V_4} = \frac{V_2}{V_3} \quad \text{or} \quad \frac{V_1}{V_2} = \frac{V_4}{V_3}$$

Hence Equation 6.5 follows:

$$\eta_C = ① - \frac{T_c}{T_h}$$

Appendix 12: Otto-cycle efficiency

Refer to the theoretical p–V diagram Figure 6.12.

$$\eta_{ind} = \frac{W_{ind}}{Q_S}$$

$$= \frac{Q_S - Q_R}{Q_S}$$

$$= 1 - \frac{Q_R}{Q_S}$$

Heat supplied, $Q_S = mc_v(T_3 - T_2)$
Heat rejected, $Q_R = mc_v(T_4 - T_1)$ (positive)

Substituting:

$$\eta_{ind} = 1 - \frac{mc_v(T_4 - T_1)}{mc_v(T_3 - T_2)}$$

$$= 1 - \frac{T_4 - T_1}{T_3 - T_2}$$

Now,

$$\frac{T_2}{T_1} = \left(\frac{V_1}{V_2}\right)^{\gamma-1}$$

and

$$\frac{T_3}{T_4} = \left(\frac{V_4}{V_3}\right)^{\gamma-1} = \left(\frac{V_1}{V_2}\right)^{\gamma-1}$$

$$\therefore \quad \frac{T_2}{T_1} = \frac{T_3}{T_4}$$

or

$$\frac{T_4}{T_1} = \frac{T_3}{T_2}$$

$$\eta_{ind} = 1 - \frac{T_1\left(\dfrac{T_4}{T_1} - 1\right)}{T_2\left(\dfrac{T_3}{T_2} - 1\right)}$$

$$= 1 - \frac{T_1}{T_2}$$

Therefore

$$\eta_{\text{ind}} = 1 - \frac{1}{\left(\dfrac{V_1}{V_2}\right)^{\gamma-1}}$$ theoretical efficiency of Otto cycle (6.6)

where V_1/V_2 is the compression ratio. Hence the efficiency of the ideal Otto cycle depends only on the compression ratio (for a given value of γ).

Appendix 13: Proof of Archimedes principle

Consider a block immersed in a fluid of constant density ρ as shown in Figure A13.

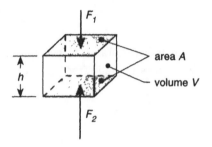

Fig. A13

The fluid pressure forces acting on the sides of the block balance and have no resultant (otherwise the block would move of its own accord sideways). However, the pressure acting on the base of the block is greater than the pressure acting on the top of the block due to the increase of pressure with depth in a fluid.

If h is the difference in height between the top and bottom surfaces, the pressure difference is given by Equation 2.8: $p = \rho g h$

Since force = pressure $\times$ area, $F = F_2 - F_1 = pA = \rho g h A$

But $\qquad hA = V$ (volume of the block)

$$\therefore F = \rho g V$$

But $\qquad \rho = \dfrac{m}{V} \; \therefore \rho V = m$

$$\therefore F = mg = w_t$$

Because ρ is the density of the fluid, w_t is the weight of the fluid displaced by the block. Therefore, the upthrust force F = weight of fluid displaced (which is Archimedes principle).

Note For simplicity, Archimedes principle was derived for a rectangular block but the shape is irrelevant and the principle is true for a block of *any* shape.

Index